NMR, NQR, EPR, AND MÖSSBAUER SPECTROSCOPY IN INORGANIC CHEMISTRY

ELLIS HORWOOD SERIES IN INORGANIC CHEMISTRY

Series Editor: J. BURGESS, Department of Chemistry, University of Leicester

Inorganic chemistry is a flourishing discipline in its own right and also plays a key role in many areas of organometallic, physical, biological, and industrial chemistry. This series is developed to reflect these various aspects of the subject from all levels of undergraduate teaching into the upper bracket of research.

NMR, NQR, EPR, AND MÖSSBAUER SPECTROSCOPY IN INORGANIC CHEMISTRY

R. V. PARISH Ph.D., D.Sc.(London), D.Sc.(Manchester), C.Chem., F.R.S.C.
Department of Chemistry
The University of Manchester Institute of Science and Technology

ELLIS HORWOOD
NEW YORK   LONDON   TORONTO   SYDNEY   TOKYO   SINGAPORE

First published in 1990 by
ELLIS HORWOOD LIMITED
Market Cross House, Cooper Street,
Chichester, West Sussex, PO19 1EB, England

A division of
Simon & Schuster International Group
A Paramount Communications Company

Typeset in Times by Ellis Horwood Limited
Printed and bound in Great Britain
by Hartnolls, Bodmin, Cornwall

British Library Cataloguing in Publication Data

Parish, R. V.
NMR, NQR, EPR and Mössbauer spectroscopy in inorganic
chemistry.
1. Inorganic compounds
I. Title
541.28
ISBN 0–13–625518–3

Library of Congress Cataloging-in-Publication Data

Parish, R. V. (Richard Vernon), 1934–
NMR, NQR, EPR, and Mössbauer spectroscopy in inorganic
chemistry / R. V. Parish
p. cm. — (Ellis Horwood series in inorganic chemistry)
ISBN 0–13–625518–3
1. Spectrum analysis. 2. Chemistry, Inorganic. I. Title.
II. Series.
QD95.P25   1990
543′.087–dc20

90–4873
CIP

Table of Contents

Preface

I should perhaps begin by explaining what this book is *not*. It is not a spectroscopic textbook, nor is it written for those with a need for detailed theory. Such books and reviews exist in abundance already. Rather, being written by a simple-minded inorganic chemist, it is intended to give an introduction to the *interpretation* of some of the types of spectra often met with in the inorganic laboratory. Theoretical aspects are kept to the bare minimum needed to allow some understanding of the methods of interpretation, and purely 'organic' applications are totally neglected. The major emphasis is therefore on 'first-order' phenomena, and more complicated cases are only briefly mentioned. Readers encountering such cases in the laboratory are strongly advised to seek advice from an expert in the technique concerned and then to delve into some of the texts mentioned in the bibliographies. For the same reason, little mention is made of the numerous double-resonance techniques now available: it is not possible to give more than the merest flavour, and it is far better to consult an expert directly, to be advised on the particular technique necessary to solve the exact problem in hand.

I am extremely grateful to many friends and colleagues who have constructively criticized various draft chapters, in particular Frank Berry, Roy Fields, Frank Mabbs, and Tony Parker, to Cliff Evans who has obtained some of the more exotic NMR spectra, and to all those who have generously permitted use of their published spectra, not to mention several generations of students who have allowed themselves to be experimented on in lecture presentations. Errors remaining are, of course, my sole responsibility, but helpful comments from readers would be welcomed.

R.V.P

1

Introduction

Chemists use a variety of spectroscopic methods to characterize and study their compounds, to follow reactions, and to understand bonding. Many textbooks present the theory of these methods, some in great detail. Few show how the chemical information may be extracted from the data, particularly for systems of interest to the inorganic chemist. The principal aim of this book is to provide guidance for the practising inorganic chemist on the interpretation of particular types of spectra and spectroscopic data, whether obtained at the bench or from the literature. Theoretical considerations are kept to the minimum required to understand the interpretative methods. Some advice on the selection and treatment of samples is also given. This chapter provides an introductory survey of the techniques discussed later in detail, and treats matters which are common to them all.

In a practical investigation, several methods are usually brought to bear on a given problem. It is seldom safe to rely on a single technique, and the greater the number of techniques employed, the greater is the likelihood of obtaining the right solution. However, each method must be understood before it can be teamed with the others, and so each will be treated separately.

The particular spectroscopic methods selected are those which deal with nuclear resonance and electron spin resonance. They are:

NMR, nuclear magnetic resonance
NQR, nuclear quadrupole resonance
NGR, nuclear gamma resonance = Mössbauer spectroscopy
and
EPR, electron paramagnetic resonance = ESR, electron spin resonance.

This particular selection has been made on the grounds that they have much in common and that all are particularly useful in inorganic chemistry.

In each case, as indeed with any form of spectroscopy, a system is probed which has various energy levels available to it. Some form of electromagnetic radiation is provided which has energy in the range needed to excite transitions between the energy levels, following the normal resonance absorption conditions $\Delta E = h\upsilon$, where h is Planck's constant 6.626×10^{-34} J s, and υ is the frequency of the radiation.

The three nuclear methods differ in the interactions which give rise to the sets of energy levels and, therefore, in the energy ranges required (Fig. 1.1). The simplest,

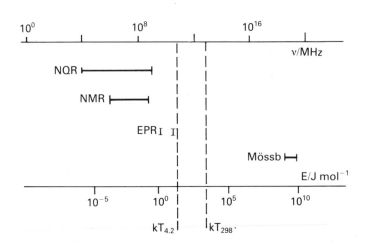

Fig. 1.1 — Energies of the exciting radiation used in various spectroscopic methods.

in principle, is NQR, in which a nucleus attains a set of closely spaced energy levels by interaction with its immediate electronic environment. Transitions between these levels are induced by the application of microwave radiation (10 kHz to 1000MHz).† The same energy levels are involved in Mössbauer spectroscopy, except that they all act as base levels for transitions to very high excited states, gained by application of gamma radiation ($10-120$ keV, 2.4×10^{12}–2.9×10^{13} MHz). In NMR, the separation of energy states is achieved by the application of a magnetic field; strong magnets are used [1.5–15 T (T = 10 kG = 10 kOe)] together with exciting radiation in the 10–600 MHz range. In all cases, nuclei in different chemical environments give discrete signals, and various types of hyperfine interaction with the surrounding electrons and with other nuclei give further information about the nature of the environment. Electron paramagnetic resonance also needs the application of a magnetic field (0.3–1.5 T) and radio-frequency radiation (9–36 GHz)†, but this time it is unpaired electrons which are probed; their interaction with the magnetic field depends on their chemical environment, and a range of hyperfine interactions can also be displayed.

1.1 HYPERFINE INTERACTIONS

It is hardly surprising that an electron-resonance method might show the effects of chemical interactions, since chemical properties are governed by electrons. On the other hand, it is not at all obvious why methods which probe the nucleus should be of

† A list of the prefixes used to denote multiples and submultiples of units is given in Table 1.1.

Table 1.1 — Common prefixes for units

Multiple	Prefix	Symbol
10^{-15}	femto	f
10^{-12}	pico	p
10^{-9}	nano	n
10^{-6}	micro	μ
10^{-3}	milli	m
10^{3}	kilo	k
10^{6}	mega	M
10^{9}	giga	G
10^{12}	tera	T

any use in a chemical context, since the nuclei lie deeply buried in the interior of their atoms. Fortunately, as indicated above, there are interactions between the electrons and the nuclei. These **hyperfine interactions** can be divided into two broad types, which may be classified crudely as direct or indirect.

The indirect hyperfine interactions are important in all the nuclear techniques described here. They operate in two different ways: *via* an electric field or a magnetic field. In the first, an unsymmetrical nucleus interacts with an unsymmetrical electric field produced by the valence electrons and the surrounding atoms or ions. The interaction gives a set of closely spaced energy levels. The separations between these levels depend on the extent of the asymmetry in the surrounding electron cloud and hence on the the structure of, and nature of bonding in, the molecule. This is the quadrupole effect, on which NQR and Mössbauer spectroscopy rely directly; it is principally an inconvenience in NMR.

The magnetic interaction is of immediate significance to NMR, since this is the only technique of the three in which an external field is routinely (and necessarily) applied to the sample. The magnetic field affects the energy of magnetic nuclei in proportion to its strength. However, the electrons exert a shielding effect, so that the magnetic field experienced by the nuclei is different from that applied. The energy of the nucleus thus depends on the electron density on its atom, allowing atoms in different chemical situations to be recognized. In the other techniques, an external magnetic field is sometimes applied in order to obtain additional information.

The direct hyperfine interactions are of importance in NMR and Mössbauer spectroscopy, and rely on the fact that some electrons spend a finite time in the same region of space as the nuclei. *s*-Electrons have a small but distinct probability of being within the nuclear volume, and of interacting with the nucleus. This affects Mössbauer spectroscopy directly, since the interaction raises the energy of the nucleus and, hence, changes the energy of the transitions. It is also important in NMR because the *s*-electrons form part of the bonding system which may connect two or more atoms; the polarization of the spins of these electrons by the nuclear spins then gives a coupling of the two nuclei which produces extra energy levels; the additional transitions provide much useful information.

Interactions between nuclei and electrons must, of course, affect both partners. It is therefore possible to see similar effects in EPR spectra, which assist the determination of structures and give information about electron delocalization through the bonds.

1.2 COMMON FEATURES OF THE SPECTRA

The discussion below is a broad-brush treatment of various features which the four techniques have in common. There are several exceptions to these generalizations which are detailed in the chapters devoted to the individual techniques.

1.2.1 Energy ranges

All spectroscopic techniques are essentially concerned with measuring energies, even though the scales may be marked in curious units such as ppm, MHz, T (or kOe), or mm s^{-1}. Strictly, none of these is a true unit of energy; they are employed for convenience (laziness). For instance, a frequency, v, should be multiplied by Planck's constant to obtain the energy, hv. Factors for interconversion and translation to conventional energy units are shown in Table 1.2. All will be used interchangeably throughout this text.

Table 1.2 — Energy-scale conversion factors

	Hz	eV	J mol^{-1}
Hz	1.000	4.136×10^{-15}	3.990×10^{-10}
eV	2.418×10^{14}	1.000	9.649×10^{4}
J mol^{-1}	2.506×10^{9}	1.036×10^{-5}	1.000

For Mössbauer spectroscopy, the energy equivalent of 1 mm s^{-1} depends on the gamma energy:
$1 \text{ mm s}^{-1} = E_\gamma/c = 3.336 \times 10^{-12} E_\gamma$.
For EPR, $\Delta v = 46.69 \, g \, c \, \Delta B \, \text{T}^{-1} \text{m}^{-1}$.

The techniques of present interest span about 16 orders of magnitude in energy of excitation, from about 4×10^{-6} J mol^{-1} (10×10^{3} Hz) to about 1×10^{10} J mol^{-1} (3×10^{19} Hz), as shown in Fig. 1.1. The large gap between Mössbauer spectroscopy and the rest is spanned by the infra-red, visible, and UV regions; these have valuable techniques associated with them, but will not be considered here.

The energy ranges shown in Fig. 1.1 represent the spread of energies between the various states of the nuclei or electrons which are being probed. The relatively small magnitudes of most of these ranges have two important implications which may affect the number of lines in the spectrum and the total intensity of the spectrum.

1.2.2 Numbers of lines

In normal practice, the sample acts as an absorber of the applied radiation. That is to say, the electrons (nuclei) are excited from the ground state to various higher states, following selection rules which differ from method to method. The number of absorption lines seen should therefore reflect a combination of the number of excited states and the selection rules. However, it often happens that the ground state is actually a set of closely spaced sub-levels. In these cases, the selection rules would usually also permit transitions from one or more of the lower-lying ones, provided that they were populated. The population of a set of energy levels follows the Boltzmann distribution, and depends on the energy separation between the levels, ΔE, and on the thermal energy, kT, which is available at the temperature of measurement,

$$N_2/N_1 = \exp(-\Delta E/kT)$$

where N_1 and N_2 are the populations of the lower and higher energy levels. Fortunately, kT is normally much greater than the widest energy spread, being about 2.5×10^3 J mol^{-1} at room temperature and 35 J mol^{-1} at the temperature of liquid helium (see Fig. 1.1). This means that all the substates have almost equal populations (see Table 1.3), and that transitions are possible from each of them. For instance, in

**Table 1.3 — Boltzmann population of a single excited state
at various energies above the ground state**

v/MHz	ΔE/J mol^{-1}	N_1/N_0 (298 K)	N_1/N_0 (4.2 K)
10	3.990×10^{-3}	0.999998	0.999886
50	1.995×10^{-2}	0.999992	0.999429
100	3.990×10^{-2}	0.999984	0.998858
500	1.995×10^{-1}	0.999919	0.994303
1 000	3.990×10^{-1}	0.999839	0.988638
5 000	1.995	0.999195	0.944468
9 000	3.591	0.998552	0.902271
36 000	14.64	0.994220	0.662747

$N_1/N_0 = \exp(-\Delta E/kT)$.
$kT = 6.210 \times 10^6$ MHz at 298 K   (2.478 kJ mol^{-1}).
$\quad = 8.751 \times 10^4$ MHz at 4.2 K   (34.92 J mol^{-1})

elemental iodine, the ^{127}I nuclei have two excited states lying 344 and 1032 MHz above the ground state, but the NQR selection rule allows transitions only between adjacent states. However, since the first excited state has a population almost equal to that of the ground state, two lines appear in the spectrum, at 344 and 688 MHz (Fig. 1.2).

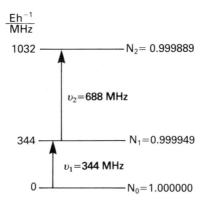

Fig. 1.2 — Energy levels, populations, and NQR transitions for I_2 at 298 K.

The situation is slightly different in Mössbauer spectroscopy, where transitions occur between states separated by about 10^{13} MHz. However, the ground state is often split by the same phenomena as occur in NQR and NMR, giving a set of effectively equally populated sub-levels, each of which can act as the starting point for transitions to the upper levels.

1.2.3 Intensities

The amount of radiation absorbed depends on the population of the starting state. Once the radiation has been absorbed, the atom (nucleus) does not normally stay in the excited state for very long. Spontaneous reversion to the lower state occurs, with re-emission of the radiation after a time delay which may be as short as a few microseconds. The excitation process, however, is several orders of magnitude faster, needing only about 10^{-17} s. De-excitation can also occur by a process known as **stimulated emission**. When radiation at the resonance frequency falls on the system, not only are nuclei (electrons) in the ground state excited, but those in the excited state are stimulated to drop down to the lower state on a similar time scale. The numbers which will jump, in either direction, are directly proportional to the populations of the states. The intensity of the absorption line, thus depends on the difference in the numbers of nuclei (electrons) making the two types of transition, i.e. on the difference in the populations. As Table 1.3 shows, these differences are very small for most of the energy ranges involved: under these conditions, the differences in Boltzmann populations are closely proportional to the energy differences. Again, Mössbauer spectroscopy is different, since the transition energies are always far greater than kT. For the other techniques, EPR gives the greatest spread of energies and, therefore, the most intense spectra. This means, in turn, that it is the most sensitive technique. Thus, while NMR samples must contain at least ca 10^{-6} mol of the nuclei being studied to give a reasonable spectrum (i.e. millimolar solutions), EPR can detect as few as 10^{11} spins, i.e. 10^{-12} mol, and will routinely give good spectra on samples containing less than a micromole for transition–metal complexes and far less for free radicals. For all the techniques, sensitivity is improved by cooling the sample, so that the population differences become greater (kT becomes smaller).

In NQR and Mössbauer spectroscopy, the spacing of the energy levels is governed entirely by the electronic conditions within the sample. The major difference between these two techniques is that the former shows only transitions between closely spaced levels (typically 10–100 MHz), while in the latter the transitions are over many eV. Consequently, it is much easier to obtain an intense Mössbauer spectrum than a strong NQR spectrum. With the other two techniques, NMR and EPR, the spacing of the energy levels depends additionally and directly on the strength of an externally applied magnetic field. The sensitivity can therefore be increased by increasing the magnetic field, and this is why NMR spectrometers with powerful superconducting magnets are so useful; they can work at 200–600 MHz, rather than 30–60 MHz for a conventional electromagnet. EPR spectrometers function at much higher frequencies (9–36 GHz) because the basic magnetic moment of an electron (the Bohr magneton) is about 2000 times times greater than that of a simple nucleus (the nuclear magneton). Their sensitivity is correspondingly greater.

A further problem which may arise from the small difference in populations of the energy states is that of **saturation**. With conventional absorption spectroscopy in, say, the visible region, where the ground and excited states are very far apart in energy, the amount of radiation absorbed depends directly on the intensity of the incident radiation. The signal-to-noise ratio can be increased by using more intense radiation. This applies to Mössbauer spectroscopy also, and it is possible because the energy gap between the initial and final states is so large that the excited state effectively has a population of zero. There is no restriction on the number of excitations possible, other than those imposed by the selection rules. For the other techniques considered here, however, increasing the radiation power can be counter-productive. This is because, in order to achieve net absorption of radiation, the rate of excitations from the ground state must exceed that of de-excitation of the upper states. This is the normal state of affairs, but, if the excitation rate becomes too great, the populations of the states rapidly become more and more equal and may even become identical. The intensity of the spectrum therefore diminishes, and may become zero. There is usually an optimum power level for each type of spectrometer, and, in principle at least, for each individual sample.

If spontaneous and stimulated emission were the only processes responsible for depopulation of the excited state, the intensity of the spectrum would slowly diminish as the spectrometer continued to run, even at the optimum power level. This does not happen because the energy accumulated by the sample can be dissipated by other means, e.g. by transfer to the lattice. Such **relaxation processes** occur considerably more rapidly than spontaneous emission. In some instances, useful information can be obtained by studying these processes, which are manifested as changes in the line widths, but their detailed treatment is beyond the scope of the present treatment. The influence of relaxation on line width is discussed below.

1.2.4 Line widths

All spectra display lines with a certain width, i.e. absorption bands are seen rather than discrete lines. Each technique has a characteristic minimum line width which is governed by the nature of the de-excitation processes involved. This occurs because the energies of the excited states are not precisely defined, but actually cover a range which depends on the time-scale of the de-excitation process. The time-scales are usually expressed as half-lives, $t_{1/2}$, i.e. the time during which a 50% reduction in population occurs, or as mean life-times, $\tau = t_{1/2} \ln 2$. Similarly, the energy range and the line width are measured as a half-width, ΔE, which is the difference between the upper and lower energies corresponding to 50% of the maximum intensity (Fig 1.3). The two are related by the uncertainty principle

$$\tau \, \Delta E = h/2\pi = 6.35 \times 10^{-11} \text{ s J mol}^{-1}$$

Thus, the longer the half-life, the more sharply defined is the energy and the narrower the absorption line. (Note that this implies, as would be expected, that the ground state has an infinite life-time, in the absence of external events.)

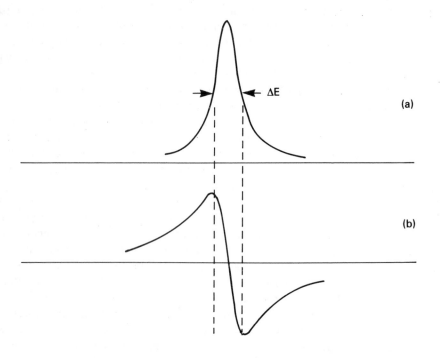

Fig. 1.3 — Definition of the half-width of an absorption line. The derivation of its correct title, full width at half-height, should be obvious. (a) is the normal intensity-spectrum, (b) the first-derivative form.

The life-times involved vary from technique to technique and sometimes more than one process may be involved. Clearly, the line width is governed by the fastest process operating. For Mössbauer spectroscopy, the life-time is essentially that for spontaneous emission, which is typically in the range 1–100 ns. Mössbauer lines are therefore usually very broad in relation to their separation. In NMR the spontaneous-emission life-time is much longer; it can be many seconds, and the spectra are correspondingly well-resolved. However, since the energies involved are now quite modest, other de-excitation mechanisms become possible. The two most common involve imparting the energy to the surrounding molecules, spin–lattice relaxation, or sharing it with other nearby spin systems, spin–spin relaxation. In dilute solutions, these are relatively slow, inefficient processes, and NMR lines remain sharp. In solids, spin–lattice relaxation usually occurs rapidly, with consequent broadening of the lines. This is why special techniques are needed to observe high-resolution NMR spectra for solids. Other factors which increase relaxation rates are the presence of paramagnetic species or quadrupolar nuclei. Electronic spin–spin relaxation in solids can be very efficient, so the EPR spectra of transition–metal compounds are often very broad and poorly resolved.

1.2.5 Types of spectrometer

With most of the techniques, the first spectrometers to be invented operated on a slow-sweep basis. The necessary energy range was covered slowly and the signal was normally displayed on a chart recorder. Resolution depended on the speed of the scan, and the signal:noise ratio largely depended on the size/concentration of the sample. Many spectrometers still work on this principle (e.g. NQR, EPR), and improvements have come mainly from reduction in noise and line width by careful design and good electronics. However, the time needed for a single scan can still be quite long, which may be a considerable disadvantage.

The signal:noise ratio can be dramatically improved if a multi-channel analyser system is used instead of a chart recorder and the signal can be converted into a series of pulses so that the number of pulses is proportional to the signal amplitude (analogue-to-digital conversion). Each channel of the analyser then contains the signal for a portion of the spectrum. If the spectrum is scanned again and the new data are added to those of the first scan, an improved total spectrum is obtained. The extent of the improvement becomes greater as the number of accumulated scans increases. This is because, on average, the desired signals go always in the same direction while the noise, being random, tends to cancel out. A simple statistical treatment shows that the signal:noise ratio increases in proportion to the square root of the number of scans. This signal-averaging method is routinely used in Mössbauer spectroscopy and continuous-wave (i.e. frequency-sweep) NMR, and EPR. It is also used in combination with the newest improvement in technique, the Fourier transform method.

As an alternative to sweeping through the entire spectral range in, say, an NMR experiment, it is possible to apply a single, very short pulse, which contains the whole range of frequencies. This changes the orientation of the spin-axes of the nuclei, so that they are now perpendicular to the magnetic field, but they precess about the field direction. Such movement of the microscopic magnets produces an alternating voltage which can be measured. The period of the alternation depends on the chemical shift. The nuclei slowly relax to the previous orientation by exchanging energy with neighbouring nuclei (spin–spin relaxation), so that the magnitude of the voltage slowly decreases. The diminishing voltage is therefore measured as a function of time, to give a free induction decay (FID) curve which contains information on the amplitude and period of the signals due to the various nuclei.

When the relaxation is complete, another pulse can be applied. The FID signals are accumulated for a given number of pulses, just as in the continuous-wave method, until the signal:noise ratio is judged satisfactory. The FID is then treated mathematically by Fourier transformation, which effectively reconverts the data to the familiar absorption versus energy form. The FT technique is now commonplace in NMR, and is beginning to be used in EPR and NQR.

1.3 QUANTUM NUMBERS

Atomic systems have to be described in terms of quantum mechanics which, for the average scientist, is not readily understandable. It has therefore become customary to make analogies between microscopic quantized systems and more familiar macroscopic ones, and to mix the quantum and classical descriptions. A similar approach will be used here.

The various energy levels involved in the transitions probed by spectroscopic methods are defined by quantum numbers. The common feature of the techniques gathered together in this book is that it is the **spin** and **magnetic quantum numbers** which are important. These numbers define the angular momenta of the nuclei or the electrons; it is these momenta which are responsible for the interactions with magnetic and electric fields. Since both nuclei and electrons are charged particles, they must interact with other electronic and nuclear charges in their neighbourhood, i.e. with the electric field due to those charges. They also behave as though they are spinning and, classically, a spinning charge must produce a magnetic field. Two electrons paired in the same orbital have opposite spin, and cancel each other out, but unpaired electrons and nuclei with non-zero spin behave like small magnets; they possess magnetic moments and will interact with magnetic fields. They also interact with each other, and all these types of interaction can be probed by spectroscopic methods. There may also be a further magnetic field associated with the orbital momentum of unpaired electrons.

For nuclei, the spin is characterized by the spin quantum number, I, which may be integral (including zero) or half-integral. The spinning nucleus appears to possess a total angular momentum of $[I(I+1)]^{1/2}h/2\pi$. The magnetic quantum number, m_I, takes values of $+I, +(I-1), +(I-2), \ldots, -(I-1), -I$. Each value of m_I corresponds to a different possible orientation of the spin axis (magnetic dipole) of the nucleus when it is in an applied magnetic field. These orientations are such that the component of angular momentum in the direction of the field is $m_I h/2\pi$ (see Fig. 1.4) Each orientation places the magnetic dipole of the nucleus at a different angle to the applied field, and therefore corresponds to a different energy for the nucleus. Thus, a nucleus with $I=1/2$ has two possible energy states, corresponding to $m_I=+1/2$ and $m_I=-1/2$. A nucleus with $I=3/2$ has four energy states, and so on. In the classical analogy, the spinning nucleus acts like a gyroscope whose axis is tilted away from the vertical. The spin axis will then rotate about the vertical axis with a precessional motion.

An electron has spin of 1/2, and thus has two possible states in a magnetic field, corresponding to $m_s=\pm1/2$ (see below for the effect of an electric field). When more than one unpaired electron is present in an atom, the unpaired spins are all parallel. The total spin is then the sum of the individual spins, and the analogous quantum

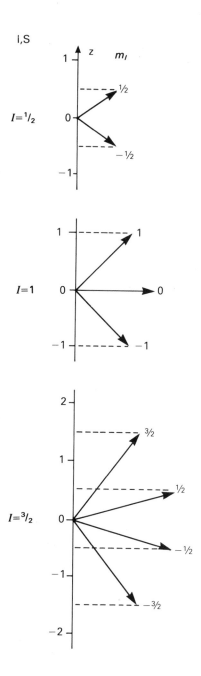

Fig. 1.4 — Quantization of nuclear spin angular momentum. The length of the arrows represent the total angular momentum, $[I(I+1)]^{1/2}h/2\pi$. For each value of I, $2I+1$ orientations are possible, each corresponding to a different m_I value, such that the z-component of the angular momentum is $m_I h/2\pi$.

numbers are S and M_S. These are related in exactly the same way as I and m_I, and sets of $2S+1$ energy levels are obtained when the atom is placed in a magnetic field.

The energy levels just described are important in NMR and EPR, in which the sample is placed between the poles of a laboratory magnet. They may also be relevant for samples in which there is an internal magnetic field produced by the alignment of electron spins on adjacent atoms. Such materials are ferromagnetic (or antiferromagnetic, when adjacent sets of spins are aligned in opposition), and the magnetic fields can be very large indeed, much greater than anything yet attainable even with a superconducting magnet.

Nuclear quadrupole resonance owes its existence to the fact that some nuclei can interact with an electric field or, more precisely, an electric-field gradient. The nuclei concerned are those with $I>1/2$. Such nuclei behave as if they possess an electric quadrupole moment, that is as if the positive charge is not distributed evenly over a sphere but is concentrated along some directions more than others. The charge distribution may resemble a rugby (or American) football, or a discus. From the symmetry of the quadrupole, there is no nett orienting interaction with an electric field provided that the field is uniform; any torque produced at one end of an axis is cancelled by that at the other end (Fig. 1.5(b)) However, if the electric field is non-uniform, i.e. there is an electric–field gradient, the two torques are not equal, and the nucleus will experience a nett force which depends on its orientation relative to the field gradient (Fig. 1.5(c)). In quantum mechanical terms the orientations, and therefore the energies, are again defined by the m_I values. However, because a quadrupole is symmetrical, pairs orientations have the same energy. It should be obvious that $\theta=t$ and $\theta=180°-t$ correspond to the same energy, since they are identical except in rotation about the field–gradient axis (compare Figs 1.5(c) and (d)). In terms of m_I, these two states have the same numerical value but opposite signs. Thus, quadrupole effects are measured by the square of m_I, and all levels are doubly degenerate (except for $m_I=0$). These effects are also seen in Mössbauer spectroscopy, where they produce additional lines in the spectra.

Similar considerations apply to electrons except that, since they possess electric dipoles, they can interact with a uniform electric field. Thus, a series of energy levels is obtained, each of which is doubly degenerate, i.e. is characterized by $\pm M_S$. This is important in EPR spectroscopy because it means that an atom containing unpaired electrons has a range of energy states available to it even in the absence of a magnetic field. (All atomic systems contain electric fields.) The **zero-field splitting** of energies is one of the factors which can influence the positions of the absorption lines.

1.4   RESTRICTIONS

All the methods have some limitations on the type of system which can be examined. EPR is restricted to samples containing unpaired electrons, principally compounds of the d- or f-block elements or free radicals. The nuclear techniques require particular isotopes containing nuclei with suitable spins and certain other properties; the elements which can be most usefully and easily examined are shown in Fig. 1.6. (Full lists are given in the Appendices.) NQR and Mössbauer spectroscopy can only be used with solid samples, whereas NMR is simple with liquids and very difficult for solids. ESR samples may need degassing. Some systems can only be studied at low

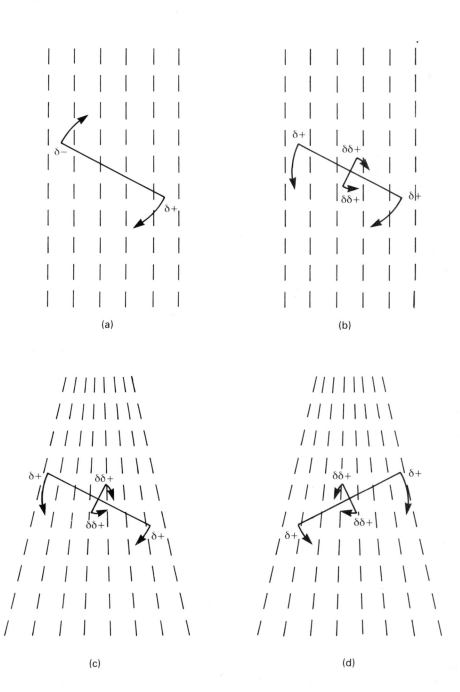

Fig. 1.5 — Interaction of a dipole and a quadrupole with uniform and non-uniform electric fields. The dotted lines represent the lines of force, and the curved arrows the torque produced by the interaction of the charge with the electric field. The dipole (a) experiences a nett orienting torque in a uniform field, but the quadrupole is oriented only in a non-uniform field (c, d).

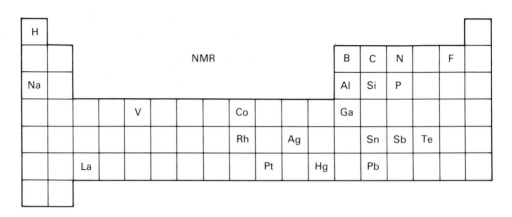

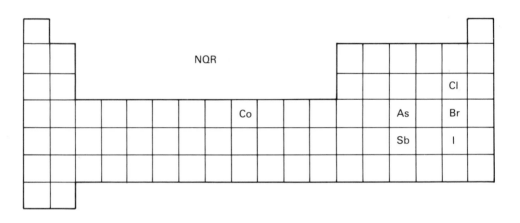

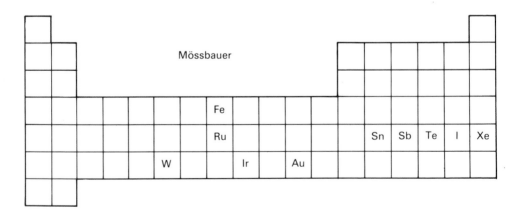

Fig. 1.6 — Elements with isotopes which can readily be examined by nuclear spectroscopic
methods. Fuller tables are given in the Appendices.

temperatures. These limitations, and the reasons for them, are explained under the individual techniques. Nonetheless, there is a wide variety of materials which can be studied easily by each technique and, in cases where more than one technique can be applied, the information obtained is usually complementary rather than repetitive. The following chapters deal with each technique in turn.

2

Nuclear magnetic resonance

The importance of nuclear magnetic resonance (NMR) spectroscopy in organic chemistry is well known. Its application has revolutionized the determination of molecular structures. The method is equally useful in inorganic chemistry; indeed, it is probably the most powerful single technique available. In addition to molecular structures, information can be obtained about the configuration and mode of bonding of ligands, molecular flexibility (fluxionality), and ligand–exchange reactions. It is sometimes possible to follow the course of reactions, and to identify the products, by carrying out the reaction in the NMR tube.

Information is given by the positions of the resonance lines and their splitting patterns, both of which reflect the chemical environment of the nucleus. The numbers of sets of lines represents the numbers of distinct chemical environments, and a good indication of the structure of a molecule can normally be obtained from its NMR spectrum alone. This process is considerably aided by the fact that it is usually possible to study more than one type of nucleus in a single molecule. Modern spectrometers are so versatile that spectra can be obtained from, say, ^{1}H, ^{13}C, ^{31}P, etc., in the ligands, and sometimes from the metal atom itself, e.g. ^{195}Pt, or ^{103}Rh. Each nucleus gives information from a different viewpoint, so it is usually quite simple to build up a picture of the whole structure. A list of the commonly studied nuclei is given in Table 2.1. A more complete list appears in Appendix 1.

Obviously, the sample must contain at least one isotope which can be probed. To obtain the most useful information, it must be likely either that the isotope occupies more than one chemically distinct position or that it will show coupling to other magnetic nuclei present in the same molecule. For instance, there is little use in studying a compound such as $[PdCl_2(Ph_3P)]_2$, in order to determine its structure, since the triphenylphosphine ligands are equivalent in all possible structures:

Table 2.1 — A selection of NMR-active nuclei

Istotope	Abundance (%)	Spin	Relative frequency (MHz)	Receptivity
^{1}H	100	1/2	100	1.00
^{11}B	80.4	3/2	32.1	4.1×10^{-2}
^{13}C	1.1	1/2	25.1	1.8×10^{-4}
^{14}N	99.6	1	10.1	1.0×10^{-3}
^{15}N	0.37	1/2	10.1	3.9×10^{-6}
^{19}F	100	1/2	94.1	8.3×10^{-1}
^{27}Al	100	5/2	26.1	2.1×10^{-1}
^{29}Si	4.7	1/2	19.9	3.7×10^{-4}
^{31}P	100	1/2	40.5	6.6×10^{-2}
^{51}V	99.8	7/2	26.3	3.8×10^{-1}
^{59}Co	100	7/2	23.6	2.8×10^{-1}
^{77}Se	7.6	1/2	19.1	5.3×10^{-4}
^{103}Rh	100	1/2	3.2	3.2×10^{-5}
^{109}Ag	48.2	1/2	4.7	4.9×10^{-5}
^{119}Sn	8.6	1/2	37.3	4.5×10^{-3}
^{121}Sb	57.3	5/2	24.0	9.3×10^{-2}
^{125}Te	7.0	1/2	31.5	2.2×10^{-3}
^{183}W	14.4	1/2	4.2	1.1×10^{-5}
^{193}Ir	62.7	3/2	1.9	2.1×10^{-5}
^{195}Pt	33.8	1/2	21.4	3.4×10^{-3}
^{199}Hg	16.3	1/2	17.9	9.8×10^{-4}

For a more complete listing, see Appendix 1.

2.I

2.II

2.III

In this case the ^{1}H, ^{13}C, and ^{31}P spectra will all yield the same information. A choice could then be made only if the chemical shift(s) were dependent on the structure and the isomers had been studied previously. However, if there is a possibility that the ligands will not be equivalent or that there may be other ligands or even the metal atom itself which can give coupling constant information (e.g. a hydride ligand, or platinum as the metal), the spectra can be very informative indeed.

Most attention will be given to ^1H spectra, because these are both the most common (especially in the less recent literature), and the most complex. The complexity comes from the number of sites which can be occupied by hydrogen atoms, particularly in organometallic compounds, but also because they all couple with each other. With other nuclei, the spectra are often artificially simplified by suppressing the coupling to ^1H.

A consideration of the experimental conditions necessary is followed by a brief, simplistic revision of the fundamentals and possible complications which may be encountered. Applications involving the commoner isotopes are then exemplified.

2.1 EXPERIMENTAL CONSIDERATIONS

In essence, an NMR experiment consists of placing a sample between the poles of an electromagnet and applying a variable radiofrequency. Excitation of the nuclei from the ground state to higher states results in absorption of radiation at particular frequencies, giving a spectrum which consists of one or more resonance lines. The number, patterns, and intensities of these lines are related to the structure of the molecule.

2.1.1 The sample

NMR measurements have to be made on liquids or solutions: useful high-resolution data for solids can only be obtained with special spectrometers, using the so-called 'magic-angle spinning' technique (MAS, see below). The compound must also be diamagnetic; paramagnetic samples usually give very broad spectra, sometimes so broad as to be undetectable, and for those which do give resolved spectra the chemical shifts (line positions) are usually anomalous. In some cases, as in organic chemistry, a paramagnetic material may be added to the solution to act as a 'shift' reagent, which may allow resolution of heavily overlapped signals by affecting the chemical shifts of some of the signals more than others, or as a relaxation agent to shorten the accumulation time or to sharpen the lines of the spectrum by decreasing the relaxation times (e.g. for ^{13}C).

Neat liquids are convenient, since the concentration is the highest possible. However, chemical shifts may not be representative, and it is often better to add an isotropic solvent. It may also be necessary to use a solvent to allow the temperature to be lowered or to reduce the viscosity: water-like mobility at all temperatures is desirable. Solid samples, of course, must be dissolved, but a reasonably high concentration is desirable ($\geqslant 10\%$), to avoid long accumulation times; concentrations less than millimolar are very difficult. (For signal-averaging or FT machines, the signal:noise ratio improves in proportion to the square root of the number of scans; scanning may be at several scans per second or several seconds per scan, depending on the isotope.) Total volumes of 0.5–2 cm^3 are required. If low temperatures are to be used, care must be taken to ensure that the sample will not crystallize out, and the solvent used must obviously remain liquid over the whole temperature range. Ideally, the solvent should not have resonances of its own within

2–3 ppm of the signal to be observed. For ^{1}H spectra deuterated solvents are often employed. However deuteration is often incomplete and ^{1}H resonances due to the solvent may be found; it is always advisable to run the spectrum of the solvent alone. One should also be aware of the possibility of H–D exchange between the solvent and the sample, e.g. via keto–enol tautomerism.

The compound used as chemical-shift standard (see Table 2.2) can often be

Table 2.2 — Compounds used as chemical-
shift standards

Isotope	Standard
^{1}H, ^{13}C, ^{29}Si	$Si(CH_3)_4$
^{14}N, ^{15}N	CH_3NO_2
^{19}F	$CFCl_3$
^{31}P	H_3PO_4 (85% aqueous solution)

added directly to the solution, provided, of course, that it will not interact with the substance under investigation. Some spectrometers allow a calibration run to be made separately, but if this is not possible the standard may be placed in a sealed capillary tube inserted in the NMR tube. Many modern spectrometers require the presence of a deuterated compound to provide a fixed signal by which the frequency may be regulated (a 'locking' signal). Most conveniently, this substance could be used as the solvent (e.g. CDCl$_3$): again care must be taken to ensure that no interaction with the solute is likely — note that many transition–metal hydride complexes react with chlorinated solvents. If the 'lock' compound is expensive or reactive, a small quantity may be sealed in a capillary or the tube containing the sample may be placed inside a slightly larger NMR tube and the 'lock' compound placed in the outer tube. This is useful if, for instance, the use of low temperatures necessitates the use of (expensive) d_2-dichloromethane (FP $-95°$C) instead of d_3-chloroform (FP $-63°$C).

2.1.2   The spectrometer

A wide variety of spectrometers is currently available, with operating frequencies ranging from less than 100 MHz to 600 MHz. Normally, the magnetic field is kept constant, and great care is taken to ensure that it is completely homogeneous over the whole of the sample. As an additional precaution, the sample tube is spun on its axis to ensure that all molecules experience the same conditions. This process can lead to the appearance of additional lines in the spectrum, which occur in pairs centred on the true resonance line and separated by the spinning frequency. Such **spinning side-bands** may be identified if necessary by changing the spinning rate.

Each spectrometer will scan a range of several kHz about the median. In general, the higher the frequency the greater the sensitivity and resolution, and the less the likelihood of second-order effects. This is not to recommend that the highest-frequency available should always be used; for many (most) purposes, a lower-frequency spectrometer is adequate, and the high-frequency machine should be reserved for special cases.

Most spectrometers have a spectrum-accumulation facility (see Chapter 1) which increases the signal:noise ratio in proportion to the square root of the number of scans. Thus, to double the signal:noise ratio, the number of scans must be increased fourfold, i.e. the accumulation time must be quadrupled. Satisfactory definition of a weak signal may therefore require many hours, and it may be necessary to change to a higher-frequency spectrometer. The increase in signal:noise ratio obtained by increasing the spectrometer frequency is due to the fact that, simultaneously, the strength of the magnet must be increased proportionately. This in turn increases the spread of energies of the nuclear orientation-states and the differences in their populations (see Chapter 1).

The **resolution** also increases with increasing spectrometer frequency. This is because the line width is governed by the physical conditions in the sample (specifically the relaxation times), and corresponds to a fixed number of hertz for a particular sample. As the spectrometer frequency is increased, the number of hertz corresponding to 1 ppm increases, so that the lines appear to get narrower (on a ppm scale). For the same reason, the chemical shift (in hertz) between resonances for two spin-coupled nuclei increases, while the magnitude of the coupling constant is unchanged; second-order effects thus become less pronounced (see section 2.2.3.2).

It is not normally possible to obtain high-resolution spectra from solid samples. This is because the lines are broadened by interactions between nuclei in neighbouring molecules, by slight differences in chemical shift owing to different orientations of the molecules relative to the magnetic field (chemical-shift anisotropy), and by slight inhomogeneities in magnetic field across the sample. For liquids, these effects are minimized by the rapid tumbling of the molecules and by spinning the sample tube. For solids, the anisotropy effects happen to become zero when the angle between the molecular and magnetic field axes is 54.7°. Thus, the sample tube is placed with its axis at this 'magic angle' to the magnetic field and is spun very rapidly so that all the interaction angles are effectively averaged to 54.7°. Even with the highest spinning rates possible, the spectra are normally rather broader than for liquid samples, so that small splittings may not be resolved, but the general principles of interpretation are unaffected. Spinning sidebands are often seen, and it can be advantageous to obtain the spectrum at two different spinning rates: the sidebands will appear at different positions but the true signals will be unaffected.

2.2 FUNDAMENTALS

In NMR spectroscopy, a liquid sample is placed between the poles of a strong magnet. Nuclei which possess magnetic moments (i.e. have spin, I, greater than zero) will tend to align themselves with the field. Other orientations of the nucleus to the field are possible, each corresponding to a different (quantized) energy, $2I+1$ in all. The actual energies are given by $-m_I\mu B/I$, where m_I is the magnetic-spin quantum number $[=I, (I-1), \ldots, -(I-1), -I]$, μ is the magnetic moment of the nucleus, and B is the magnetic field which the nucleus experiences (in general this is different from the applied field because of the shielding effects of the electrons). The total energy range covered, $\pm\mu B/I$, thus depends on the strength of the magnetic field and on the spin and the magnetic moment of the nucleus. This range is very small compared with the thermal energy available (kT). Even for the most powerful magnets currently employed, the difference between the lowest and the highest energies is less than 0.2 J mol^{-1} (see Chapter 1). This means that all the possible orientations occur with essentially equal probability. The nucleus can be excited from one orientation to a higher-energy orientation by application of radiation of appropriate frequency: conventional magnets have fields of 1–15 T (1 T=10 kOe) and frequencies of 10–600 MHz are used for the common nuclei. The selection rule is $\Delta m_I=1$, so that all $2I$ possible transitions have the same energy, and each distinct nucleus gives only a single resonance line in the spectrum. The position of that line depends on the shielding effects of the surrounding electrons which reflect the chemical environment of the nucleus. The line position is therefore known as its **chemical shift**, and is measured relative to a suitable standard substance.

The spread of chemical shifts for any one isotope is usually quite small, typically 3–30 kHz. These ranges are far smaller than the differences in $\mu B/I$ so that, in general, the frequencies for different isotopes do not overlap. Therefore, a probe which can be tuned to an appropriate range of frequencies is required for each different type of nucleus.

When the isotope being examined has other magnetic nuclei in its vicinity, those nuclei will often modify the magnetic field experienced by the nucleus in such a way as to split the resonance line. Such effects are known as **spin–spin coupling**, and the pattern and separation of the component lines reflects the numbers of adjacent magnetic nuclei and the nature of the bonds connecting them.

It should be noted that it is often difficult to obtain spectra for isotopes with $I>1/2$. The essential reason for this is that such nuclei are quadrupolar (see Chapter 3), and interact with an unsymmetrical electric field. This interaction is comparable in magnitude to that between the nucleus and the magnetic field. Since the two interactions operate simultaneously, the nucleus is able to exchange energy betwen them. Thus, a nucleus which has been excited by the NMR frequency can de-excite by transferring energy to the quadrupolar system. This energy transfer is often extremely efficient, and allows very rapid exchange of energy. Such **quadrupolar relaxation** results in considerable broadening of the resonance lines, which may be so severe as to prevent observation of the line. It is to be expected whenever the *local* symmetry of the quadrupolar nucleus is lower than O_h or T_d (see Chapter 3); this phenomenon is therefore very common.

2.2.1   Chemical shift
Every nucleus is surrounded by an electron cloud which exerts a shielding effect, so that nuclei in different chemical environments experience different magnetic fields and resonate at slightly different frequencies. It is this phenomenon which gives the diagnostic value of the technique, and characteristic ranges can be associated with particular chemical groups.

Since each type of nucleus has its own frequency range (for a given magnetic field strength), it is not necessary to know the absolute energies involved; signal positions are usually expressed in parts per million of the basic frequency of the probe in use.

These positions are referred to as chemical shifts, and are measured relative to the position of the signal for a standard compound, e.g. Me_4Si (TMS) for ^1H, ^{13}C, and ^{29}Si (see Table 2.1). The chemical shift is usually given the symbol δ,

$$\delta/ppm = 10^6 \text{ (shift/Hz)/(spectrometer frequency/Hz)}$$

The chemical shift for a particular nucleus in a given compound thus always has the same value, whatever spectrometer frequency is used.

The modern convention is to take the chemical shift as positive when it is to high frequency of the standard. This situation may also be referred to as a shift to low field. Older literature may have values with different conventions. In particular, ^1H spectra are often given on the τ-scale, for which the standard (TMS) is assigned the value 10 τ, and the direction of positive shift is to low frequency (high field). For ^1H, the convention which is being used is usually obvious, since data are quoted in units either of τ or of δ. Care must be taken with other nuclei, especially ^{31}P data quoted in the 1970s or earlier. However, the older convention is retained at least as far as plotting the spectrum is concerned: the highest frequency is at the left-hand side.

The convention to assign positive or negative chemical shifts in terms of increase or decrease in frequency follows from the fact that, in scanning the spectrum, it is normally the radio-frequency which is varied while the magnetic field is kept fixed. The fact that the frequency- and field-scales work in opposite directions sometimes causes confusion. The reason can be seen as follows. The chemical shift reflects the shielding which the nucleus experiences by virtue of the surrounding electrons. The magnetic field at the nucleus is therefore less than that applied. Hence, to achieve resonance and give a signal in the spectrometer, either the radio-frequency can be decreased until the resonance frequency at this (reduced) magnetic field is reached (a shift to low frequency), or the frequency could be kept fixed, in which case resonance could only be achieved by increasing the applied magnetic field (a high-field shift).

2.2.2 Coupling constants

In addition to being sensitive to the surrounding electron density, the nuclear resonance is also affected by the presence of neighbouring magnetic nuclei. Each such nucleus acts as an additional magnet whose effect is added to that of the spectrometer magnet. For a neighbour nucleus of spin I, each of the $2I+1$ orientations corresponds to a different field increment, so that the nucleus probed is equally likely to experience any one of $2I+1$ different magnetic fields, giving a set of signals of slightly different frequencies. Thus, a nucleus close to one other nucleus of spin I will give a signal consisting of $2I+1$ equally intense components. For example, the methyl protons of an $H_3C.CH$ group will give a doublet signal because of the single ^1H nucleus adjacent. Note that any nucleus of spin 1/2 produces the same effect, so that the hydride ligand in a rhodium–hydride complex such as $[RhHCl_2(AsPh_2Me)_3]$ also gives a doublet spectrum because of splitting by the metal nucleus [^{103}Rh, 100%, $I=1/2$]. Similarly, the ^{13}C signal of $CDCl_3$ is a 1:1:1 triplet (^2D, $I=1$). The spacing between the components of these signals is referred to as the spin–spin coupling constant, since it represents the energy of the magnetic coupling between the two nuclei. The magnitude of the coupling constant depends on the identity of the two nuclei and on the nature and number of the bonds connecting them and, hence, on the stereochemistry.

Coupling constants are always quoted in frequency units (Hz), which are independent of the frequency of the spectrometer. However, when the same coupling pattern is observed on different spectrometers, the individual components will appear at different positions on the chemical shift scale, because 1 ppm corresponds to different numbers of hertz (one-millionth of the spectrometer frequency). Coupling constants are usually given the symbol J, and the two nuclei which are coupled are indicated, e.g. J(H–P). Since the value of J depends on the number of bonds separating the two nuclei, this number is usually given as a superscript, e.g. for the cases mentioned above, ^{3}J(H–H) is about 7 Hz in the *iso*-propyl group, ^{1}J(H–Rh) is 4 or 9 Hz, depending on the stereochemistry of the complex, and ^{1}J(C–D) is 32 Hz in $CDCl_3$.

The strength of the coupling also depends on the properties of the two nuclei and of the intervening electrons. Spin–spin coupling is usually considered to be due mainly to the Fermi contact mechanism (since this can be treated theoretically relatively easily). That is, the two nuclei are in contact with the electrons of the bond(s) connecting them, which provide the coupling. Since only *s*-electrons have finite probability of being at the nucleus, the coupling constant depends on this probability and the extent of *s*-orbital participation in the bonding. Thus, the Fermi contact term for a one-bond coupling has the form $F_1 F_2 \Delta E$, where $F_1 = \gamma_1 |\psi(0)|_1^2 \alpha_1^2$, γ_1 is the magnetogyric ratio of nucleus 1, $|\psi(0)|_1^2$ is the *s*-electron density at that nucleus and α_1^2 is the *s*-character of the hybrid used by atom 1 in the bonding orbital. F_2 is defined similarly. ΔE is an electronic energy term. In series of analogous compounds, the terms other than γ are roughly constant, and estimates of coupling constants can be made by scaling the value for one compound by the ratio of γ-values. The closer the similarity between the compounds, the better this will work, e.g. ^{1}J(C–H) in $CHCl_3$ could be estimated from ^{1}J(C–D) in $CDCl_3$ as $(32 \text{ Hz}) \gamma(^1\text{H}) / \gamma(^2\text{H}) = 208$ Hz — the observed value is 209 Hz.

When coupling involves a transition-metal nucleus, this type of comparison works fairly well for related compounds of the same transition series. However, the covalency of the metal–ligand bond increases from 3*d*- to 4*d*- to 5*d*-metals, and the coupling constants are correspondingly larger. For instance, ^{1}J(M–H) and ^{1}J(M–P) differ by factors of 50–100 and 15–20 for M = ^{103}Rh and ^{195}Pt, with platinum giving the larger J-values, although the ratio of γ-values is only 6.8.

The coupling arises because the distribution of the *s*-electrons is influenced by the spin of the nucleus; there is a slightly greater probability for the electron of spin parallel to that of the nucleus to be in the vicinity of the nucleus. This polarization of spin in the bond is reversed at the second nucleus, which can thus sense the orientation of the first nucleus. The value of the coupling constant must therefore be the same for both nuclei. Thus, if the ^{2}D spectrum of $CDCl_3$ were measured, a coupling of 32 Hz due to ^{13}C (1%, $I = 1/2$) would be seen. Similarly, the ^{103}Rh spectrum of the hydride complex would show a doublet splitting of 4 or 9 Hz, and the signal for the –CH proton of the CH_3CH group should show a coupling of about 7 Hz due to the methyl protons. However, the fact that there are three such protons introduces a further complication.

Each methyl proton can have one of two possible orientations, ↑ or ↓, so that four combinations are possible, ↑↑↑, ↑↑↓, ↑↓↓, and ↓↓↓. Each combination corresponds to a different total magnetic field, but they differ by the same amount. The –CH– proton will experience four different magnetic fields, and will require a different frequency for each, giving a set of four equally spaced signals. However, the signals are not equally intense, but have the ratio 1:3:3:1. This happens because, although ↑↑↑ and ↓↓↓ are unique, the magnetic field corresponding to ↑↑↓ can also be obtained from ↑↓↑ and ↓↑↑, and that for ↑↓↓ from ↓↑↓ or ↓↓↑. All eight arrangements are equally probable, because the energy differences between them are so small.

Similar arguments show that a set of n equivalent adjacent nuclei with $I=1/2$ will result in a splitting of the signal into $2n+1$ components, equally spaced by an amount equal to the coupling constant. The intensities are in the ratios of the binomial coefficients, but are perhaps more easily derived by the J-tree technique, in which the effects of each individual nucleus are considered consecutively. Thus, a single spin–1/2 nucleus produces a 1:1 doublet with a separation of J (Fig. 2.1(a)). A second such

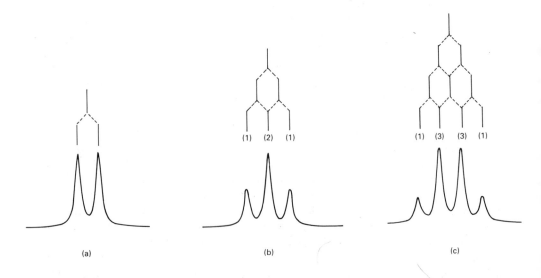

Fig. 2.1 — Characteristic coupling patterns due to (a) one, (b) two, or (c) three equivalent $I=1/2$ nuclei. The numbers in brackets are the relative intensities (numbers of superposed lines).

nucleus will split each of the doublet lines into two, again separated by J, so that the two central lines superpose, giving a 1:2:1 triplet. This process is continued until all the coupling nuclei have been included (Fig. 2.1(b)). Note that equivalent nuclei appear not to couple with each other; this is actually an extreme example of second-order effects giving unexpectedly simple spectra (see section 2.2.3.2).

It often happens that the coupling nuclei are not all equivalent. Nevertheless, it is still possible to use the same procedure, provided that the appropriate coupling constants are used. Thus, for coupling to two non-equivalent nuclei (each with $I=1/2$), two different J-values are involved, and a $1:1:1:1$ four-line spectrum is expected (Fig. 2.2). Note that the same result is obtained regardless of the order in

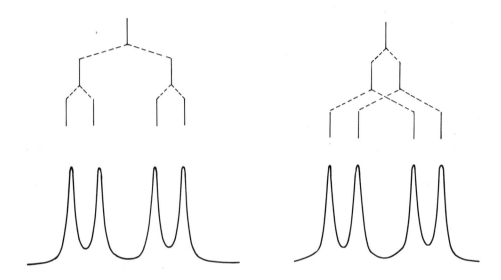

Fig. 2.2 — Coupling to two non-equivalent $I=1/2$ nuclei. Note that the same pattern is obtained whether the larger or the smaller coupling is considered first.

which the couplings are considered. Greater clarity in the diagram is usually obtained by taking them in decreasing magnitude. This scheme may readily be extended to more complex cases, for example, when the nuclei have $I>1/2$ (see Fig. 2.3). It should also be said that coupling to nuclei with $I>1/2$ is often not observed. This is because such nuclei possess quadrupole moments which interact with an electric-field gradient owing to an unsymmetrical arrangement of electrons (see Chapter 3). The interaction is comparable with or stronger than that between the nucleus and the magnetic field. As the molecule tumbles about in the solution, the quadrupolar nucleus tends to tumble with it. Its orientation with respect to the nuclei under observation is thus completely randomized and the spin–spin coupling is averaged to zero. This is always the case with, for instance, halogen nuclei (other than ^{19}F), and often applies to other nuclei such as ^{14}N. If the quadrupolar nucleus is at a site of high symmetry (strictly octahedral or tetrahedral, O_h or T_d), the electric-field gradient is zero, the nuclear reorientation does not occur, and spin–spin coupling will be seen. Thus, the ^1H spectrum of the ammonium ion is a well-resolved $1:1:1$ triplet. When the quadrupolar coupling is weak but not zero, some line broadening may be seen.

In interpreting spectra, it is usually best to ignore the quadrupolar nuclei in the first instance, considering only first-order coupling to spin–1/2 nuclei. Any features which remain unexplained may then be due to other nuclei, second-order coupling, relaxation, etc. Some of these effects are considered in the next section.

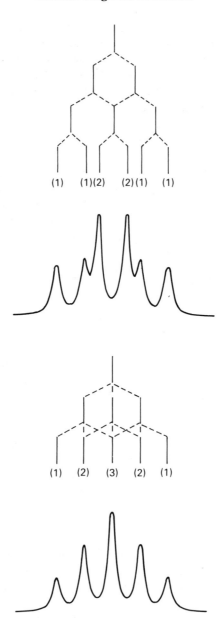

Fig. 2.3 — (a) Coupling to three $I=1/2$ nuclei, two of which are equivalent, giving a triplet of doublets. (b) Coupling to two equivalent $I=3/2$ nuclei. Figures in brackets are the relative intensities.

2.2.3 Complications
2.2.3.1 Satellites
In the examples of coupling considered above, it was assumed that all the coupling

nuclei had abundances of 100%. This is not always the case: some elements exist as mixtures of isotopes, each of which has its own spin, magnetic moment, etc. Under these circumstances, additional signals may appear. For example, natural tin consists predominantly of isotopes with zero spin (^{116}Sn, 15%; ^{118}Sn, 24%; ^{120}Sn, 32%; ^{122}Sn, 4.6%, ^{122}Sn, 5.7%, etc., a total of 83.7%). Thus, the majority of molecules of, say tetramethyltin, $(H_3C)_4Sn$, contain tin nuclei with no spin, and their ^1H NMR spectra are identical singlets (since all twelve methyl protons are equivalent). However, two isotopes of tin have spin of 1/2: ^{117}Sn (7.7%), and ^{119}Sn (8.6%). Molecules containing these isotopes will give spectra which are doublets owing to the hydrogen–tin coupling, with intensities corresponding to the abundances. Each isotope gives a different coupling constant: $^2J(^{117}Sn–H)=52$ Hz, $^2J(^{119}Sn–H)=54$ Hz. For each component of a doublet, the total intensity is divided equally between the two lines, and the final spectrum is the sum of all the individual component spectra. The main singlet thus represents 83.7% of the total intensity, and is flanked by the doublets, giving a 4.3:3.8:83.7:3.8:4.3 pattern (see Fig. 2.4). The outlying lines are known as **satellites**.

This feature is not uncommon. In principle, all ^1H spectra of organic compounds should show satellites due to ^{13}C (1%, $I=1/2$) but, since each satellite signal has only 0.5% of the total intensity, they are usually not observed. A particularly striking case is platinum, where one third of the nuclei have spin–1/2; in the spectra of platinum compounds, each satellite has 16.7% of the total intensity (see section 2.3.3.5).

The phenomenon is, of course, not restricted to ^1H spectra, but applies to any nucleus in a similar situation. The ^{13}C spectrum of tetramethyltin shown in Fig. 2.4(c) has the same pattern as the ^1H spectrum but with different values for the coupling constants.

Satellites can be recognized from the fact that they lie symmetrically about the main signal, but their intensities are not the simple fractions required for normal coupling patterns. The satellites often (but by no means always) show the same coupling pattern as the main signal. In some cases, however, the satellite patterns may not be fully resolved from the main signal, e.g. in the ^1H spectra of organophosphines bound to platinum.

2.2.3.2 Second-order spectra

The description of the effects of spin–spin coupling given above is simple and straightforward. It is known as the **first-order** treatment, and its success depends on there being a large energy separation (chemical shift) between the resonances of the individual nuclei. When the coupled nuclei belong to different isotopes, e.g. ^1H and ^{31}P, there is no problem since the chemical shifts occur in grossly different frequency regions. As the two signals approach each other, the coupling pattern begins to distort; extra lines may appear and eventually, the spectrum is completely different from the simple first-order pattern. A much more complex, **second-order** treatment is now required; the full treatment is beyond the scope of this book, but is well described elsewhere (see Bibliography). Second-order effects become marked when the difference in chemical shift between the coupled nuclei is less than 4–5 times the coupling constant. Obviously, in order to make this comparison, the chemical shift difference and the coupling constant must both be measured in the same units, e.g. hertz. The frequency corresponding to a given chemical-shift difference is directly

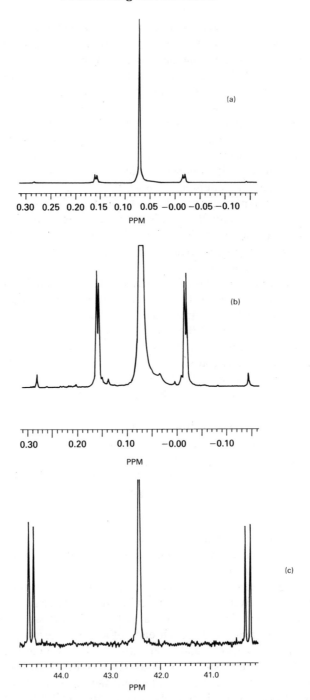

Fig. 2.4 — Satellites due to ^{117}Sn and ^{119}Sn in the spectrum of Sn(CH$_3$)$_4$; (a) the ^{1}H spectrum at 300 MHz, (b) the ^{1}H spectrum on an expanded intensity scale, (c) the expanded ^{13}C-{^{1}H} spectrum at 75 MHz. ^{2}J(^{117}Sn−H)=52 Hz; ^{2}J(^{119}Sn−H)=54 Hz; ^{1}J(^{117}Sn−^{13}C)=317 Hz; ^{1}J(^{119}Sn−^{13}C)=329 Hz.G.

proportional to the spectrometer frequency, while coupling constants always correspond to the same (constant) frequency. Therefore, second-order effects can sometimes be reduced by using a higher-frequency machine. For instance, in a 100 MHz ^{1}H spectrum, a 10 Hz coupling would probably show second-order effects when the signals for the coupled nuclei were less than about 0.5 ppm apart (see below).

In referring to spectra of this type in a general way, it is conventional to represent the nuclei by letters of the alphabet. The separation in chemical shift is then indicated by the alphabetical separation. For example, two nuclei with well-separated resonances would be labelled A and X (first-order coupling), whereas two with very close resonances would be labelled A and B (second-order).

A frequently encountered type of second-order effect is found in the ^{1}H spectra of organic derivatives of main-group elements. For example, tetraethoxysilane, $(H_3CCH_2O)_4Si$, gives a well-resolved, first-order A_3X_2 pattern. The methyl group appears as a triplet centred at 1.20 δ and the methylene group as a quartet centred at 3.85 δ, each with a coupling constant of 7 Hz (Fig. 2.5(a)). However, the spectrum of tetraethylsilane, $(H_3CCH_2)_4Si$ is completely different. The chemical shift of the CH_2 group is now much closer to that of the CH_3 group, and a complex, second-order A_3B_2 pattern is observed (Fig. 2.5(b)).

When the chemical-shift difference becomes zero, and the nuclei are identical, as the three protons of the methyl group, they appear not to couple with each other. This is actually a special case of second-order coupling, in which the intensities of all the lines other than those at the central position are zero.

| 2.IV | 2.V | 2.VI | 2.VII |

A more difficult case to recognize is that of **magnetic non-equivalence**. This is probably best described by means of examples. Consider the square-planar metal complexes **2.IV–2.VII**, in which the M, L, and L$'$ nuclei are non-magnetic. Coupling should be observed between the ^{1}H and ^{31}P nuclei. The first case (structure **2.IV**) shows straightforward AX coupling, and the ^{1}H and ^{31}P spectra will each consist of a doublet with the same coupling constant, $^2J(H–P)$. The ^{1}H spectrum for structure **2.V** would be a first-order, AXY, doublet of doublets pattern, showing different coupling constants between the proton and the two non-equivalent phosphorus nuclei, $^2J(H–P_{cis})<^2J(H–P_{trans})$. For the third structure, **2.VI**, it is evident that the two protons are non-equivalent, and two AX doublets should be seen. (The H–H coupling is very small, and probably will not be observed.) In structure **2.VII**, the two protons are chemically identical; they are related by symmetry, as are the two phosphorus nuclei. One might expect the ^{1}H spectrum to be the same as for **2.V** but with double the intensity. This is not what is observed. The two protons are chemically and symmetrically equivalent, but they are not magnetically equivalent. This is because they do not couple equally to a given phosphorus nucleus; with

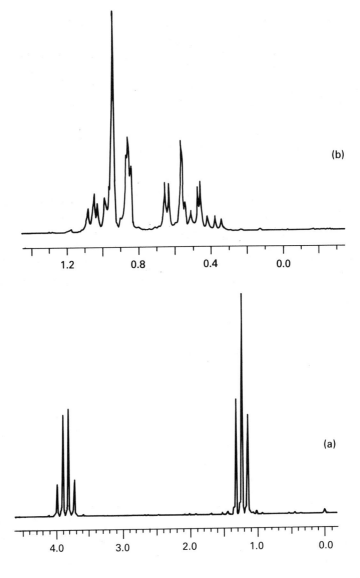

Fig. 2.5 — The ^{1}H spectra of (a) Si(OC$_2$H$_5$)$_4$ and (b) Si(C$_2$H$_5$)$_4$ at 80 MHz. Satellites due to ^{29}Si are just visible.

respect to that phosphorus nucleus, the protons are non-equivalent. Similarly the two phosphorus nuclei are non-equivalent. That is,

$$^2J(H_a\text{–}P_a)=^2J(H_b\text{–}P_b) \quad \text{and} \quad ^2J(H_a\text{–}P_b)=^2J(H_b\text{–}P_a)$$

but

$$^2J(H_a\text{–}P_a)\neq^2J(H_b\text{–}P_a) \quad \text{and} \quad ^2J(H_a\text{–}P_b)\neq^2J(H_b\text{–}P_b)$$

Under these conditions (AA'XX'), the spectra can be quite complex, and may require computer fitting or simulation for their solution.

Occasionally, however, second-order spectra may be misleadingly simple. A good example, dealt with in section 2.3.2 is the $A_nXMX'A_n'$ system, e.g. when A and A' are methyl protons and X, X' are phosphorus nuclei. In principle, the A spectrum should consist of $2n+1$ pairs of lines placed symmetrically about the chemical shift position. However, it sometimes happens that the lines cluster together and are not fully resolved. The spectrum may then appear to be a simple doublet or triplet, depending on the magnitudes of the coupling constants.

2.2.3.3 *Exchange*

Another effect which can change the shape of a spectrum is **exchange**. That is, two or more atoms are apparently able to change places and become equivalent when they would otherwise be expected to be non-equivalent. For example, the ring protons of an η^5-cyclopentadienyl group usually appear equivalent, and give a single resonance, even when the molecular symmetry is low, as in $[(\eta^5\text{-}C_5H_5)Fe(CO)_2]_2$ (**2.VIII**) or $(\eta^5\text{-}C_5H_5)_2TiCl_2$ (**2.IX**). In both of these cases, the molecule has a plane of symmetry

2.VIII   2.IX

which bisects the five-membered rings, and one would expect to see three ^{1}H resonances in ratio 1:2:2. In fact, each ring is rotating rapidly in its own plane, so that all the protons have the same average environment and become indistinguishable.

2.X

A similar effect is seen in the trimethylaluminium dimer (**2.X**), in which rapid exchange between the terminal and bridging positions renders all the methyl groups equivalent. At room temperature, a single average signal is seen, at -0.3 δ. When the sample is cooled, the exchange slows down; the spectrum begins to broaden, then to split into two broad signals, and eventually becomes two sharp resonances at -0.6 δ (intensity 4, terminal) and $+0.5$ δ (intensity 2, bridging). Note that the high-temperature spectrum is the *weighted* average of the resolved signals. This type of behaviour is also often shown by transition–metal complexes which contain several hydride ligands which undergo positional scrambling at room temperature. Molecules of these types are said to be **fluxional**, and the spectrum is affected when the rate of exchange becomes comparable with the spectrometer frequency. Note that, if spin–spin coupling is present, it also becomes averaged.

Similar effects can be produced by rapid intermolecular exchange or rapid reversible dissociation. For instance, many transition–metal complexes of tertiary phosphines undergo rapid ligand–exchange in the presence of the free ligand, so that only a single ^{31}P resonance is seen in the high-temperature spectrum. The occurrence of this type of exchange can be recognized if the coordinated phosphine is expected to show spin–spin coupling to some other nucleus in the complex, e.g. a hydride ligand or the metal itself: dissociative exchange will destroy the coupling.

The effects described above become important when the rate of exchange is appreciably greater than the characteristic frequency of the spectrometer. When the two are comparable, the spectrum may show some of the features expected for either the rapid- or the slow-exchange situation, but with greatly broadened lines. In this case, closer approach to the true, slow-exchange spectrum may be obtained by using a spectrometer with a higher operating frequency. It is possible to simulate and to analyse the line shapes expected for various rates of exchange, and thus obtain information about the rates, molecular dynamics, etc. (see Bibliography).

2.3 ^{1}H SPECTRA

NMR spectra for ^{1}H are usually called proton spectra, even when the proton is regarded, chemically, as a 'hydride'. They are the most readily available, and may be very simple or quite complex. The latter feature arises from the effects of magnetic coupling which may be to other protons or to nuclei such as ^{31}P in phosphine ligands or metals such as ^{119}Sn, ^{195}Pt or ^{103}Rh. The structure of the spectrum provides much useful information.

Two broad classes of inorganic compound may be distinguished: those which contain organic groups bound directly to an atom other than carbon (organometallics) and those which contain individual hydrogen atoms bound to atoms other than carbon (hydrides).

2.3.1 Organometallics

Organometallic compounds (which may include boron, silicon, etc., as the 'metal') usually give ^{1}H spectra very similar to those of conventional organic compounds containing the same groups. Since organic applications of NMR are well described elsewhere (see Bibliography), it is not proposed to give a detailed treatment here. For convenience, the conventional chemical-shift ranges for organic groups are shown in Fig. 2.6. The major differences between organometallics and purely organic compounds are in the chemical shifts of some groups, the occurrence of additional coupling, and fluxional effects.

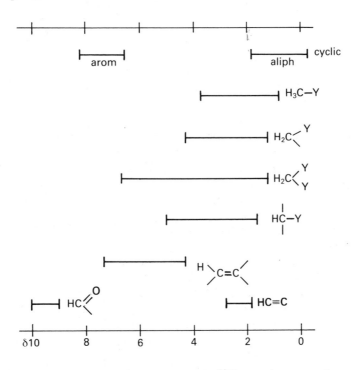

Fig. 2.6 — Characteristic chemical-shift ranges for ¹H in organic compounds.

2.3.1.1 Chemical shifts

Except for the group bonded directly to the metal atom, the chemical shifts of organic groups in organometallic compounds fall in the conventional regions. The protons on the group immediately adjacent to the metal atom often undergo a shift to low frequency of 1–2 ppm. Thus, methyl groups may be found with chemical shifts down to about −2 ppm (see Table 2.3). Normally this causes little confusion, and methyl-group signals are readily identified. Methylene groups can be similarly affected, so that their chemical shifts become comparable with those for conven-tional methyl groups. This means that ethyl groups in organometallics often give complicated second-order patterns (see section 2.2.3.2 above) rather than the normal simple quartet-plus-triplet.

σ-Bonded vinyl and allyl groups mainly give spectra of the expected shapes (Table 2.4), except that many main-group and some transition-metal allyls undergo a rapid exchange which renders the two CH_2 groups equivalent, giving an AX_4 spectrum.

Table 2.3 — Typical ^{1}H chemical shift ranges for
methyl groups in organometallics

Bonded to	δ		
C	0.7	to	1.8
Li, Be, Mg	−1	to	2
Al–Tl	−0.5	to	1.5
B, Si–Pb, P–Bi	0	to	1.5
Zn–Hg	−0.5	to	3.2
Transition metals	0	to	1.5

Precise values within these ranges depend primarily on the
electronegativity of the other groups bonded to the metal.
Thus, metal-halide derivatives fall at the lower end of the
range, polyorganic derivatives at the upper end.

Table 2.4 — ^{1}H chemical shifts (δ) in σ-bonded vinyl and allyl derivatives

	M=Main group Metal	Transition Metal	M=Transition metal
$\delta(H^a)$	6–7	8–8.5	2.0–2.5
$\delta(H^b)$ (*trans*)	5.5–6.5	2.5–3.0	6.0–6.5
$\delta(H^c)$ (*cis*)	5.0–6.0	3.5–4.0	4.5–5.0

Anomalous chemical shifts are also found for the protons of π-bonded molecules,
i.e. alkenes, alkynes, and aromatics, which all show more or less pronounced shifts to
lower δ-values. The magnitude of the shift depends on the metal and on the other
ligands, and can be quite substantial (Table 2.5). For instance, the vinylic protons of
the two double bonds in Ir(O$_2$CCH$_3$)(cod) (**2.XI**, where cod is 1,5-cyclo-octadiene)
have a chemical shift of δ 4.00 while the product of reaction with triphenylphosphine
(**2.XII**) shows resonances at δ 2.79 (*trans* to the acetate) and δ 5.06 (*trans* to the
phosphine).

2.XI   2.XII

Table 2.5 — ¹H chemical shifts (δ) in π-bonded organometallics

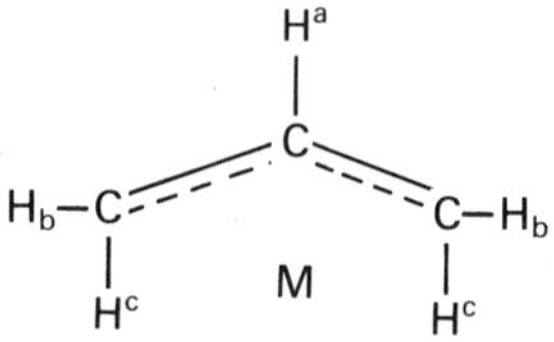

		η^5-C_5H_5	η^6-C_6H_6
H^a	4–7		
H^b	2.5–4.5	4–6	4–6
H^c	0.5–3.5		

π-Bonded groups also often show various types of fluxional behaviour, so that the spectrum does not at first sight appear to be consistent with the molecular structure. For instance, as mentioned above, many σ-bonded allyl groups undergo a rapid σ–π exchange which renders the CH_2 groups equivalent. The majority of π-bonded cyclopentadienyl or benzene complexes show only a single ¹H (or ¹³C) resonance, even when the symmetry of the whole molecule is very low. This is because the rings are able to rotate very rapidly in their own plane, so that all hydrogen (and carbon) atoms become equivalent on the NMR time-scale. So facile is this process that it often occurs even in solid materials, and sharp ¹H(¹³C) spectra may be obtained from powdered samples.

More detailed treatments of ¹H spectra for organometallic compounds may be found in the works cited in the Bibliography.

2.3.1.2 Coupling effects

Protons in organometallic compounds may couple with other protons and/or with other active nuclei. In the majority of cases, proton–proton coupling gives the pattern expected for the group concerned, the major exceptions being those mentioned in the last section. Coupling with other nuclei in the system is quite common. In instances where the coupling nuclei have 100% abundance, the coupling should be simple to recognize, since the multiplicity of the splitting pattern is simply increased. Two common examples are ³¹P and ¹⁰³Rh, both of which have spin 1/2. However, as illustrated above, the extra coupling may be obscured if the proton resonance is part of a second-order system. Also fairly common is the case where the coupling atoms have less than 100% abundance, e.g. for ^{117}Sn, ^{119}Sn, ^{195}Pt. The additional coupling now affects only the fraction of molecules which contain these isotopes, and satellite signals are seen on either side of the main signal. The case of tin compounds was dealt with in section 2.2.3.1 above. For platinum, the isotope ^{195}Pt has 33% abundance and spin 1/2, while all other isotopes have zero spin. The satellites thus represent 33% of the total intensity, which is divided equally between the two halves of the doublet. The whole pattern thus resembles a triplet, but has a 1:4:1 intensity pattern (see Fig. 2.13 in section 2.3.3.5). The coupling constant, J(Pt–H) is the separation between the two lines of intensity 1. The values of the coupling constants depend not only the nature of the organic group, but also on the numbers of bonds between the coupling nuclei and on the identity of the nuclei. For instance, coupling to ^{103}Rh is much weaker than to ^{195}Pt (because of differences in magnetogyric ratios of the two nuclei and in bond-covalency): for CH$_3$–M, ^{2}J(M–H) is 50–80 Hz for ^{195}Pt but only 3–9 Hz for ^{103}Rh.

2.3.2 Phosphine and arsine ligands

Tertiary phosphines and arsines are common ligands in transition-metal chemistry. Methyldiphenyl and dimethylphenyl derivatives are particularly useful, not only because they may confer a convenient degree of solubility, but also because their ^{1}H NMR spectra are simple and have good diagnostic value. The more sterically demanding *t*-butyl analogues have also become popular recently. For the arsines, the number and intensities of the methyl or *t*-butyl signals give a simple measure of the numbers of non-equivalent ligands present. The spectra of phosphine ligands show additional effects due to coupling with ^{31}P. Their utility is illustrated here by consideration of the phosphine PMe$_2$Ph. It should be noted that, although ^{75}As has $I=3/2$, organo-arsines and their complexes do not show ^{1}H–^{75}As coupling because of quadrupolar relaxation (see section 2.2.2).

The ^{1}H spectrum of PMe$_2$Ph shows a complex resonance at about 7.5 δ due to the phenyl group and a sharp doublet at 1.39 δ (in CDCl$_3$) owing to the methyl groups. The phenyl signal is usually complex and cannot easily be analysed (it contains lines due to the *ortho*-, *meta*- and *para*-hydrogen nuclei, which couple together and with the phosphorus nucleus). The splitting of the methyl signal is, of course, caused by coupling with the phosphorus nucleus, ^{2}J(P–H)=1.7 Hz. When the phosphine co-ordinates to an acceptor, the chemical shift changes slightly but ^{2}J(P–H) increases dramatically, so that the doublet is now well resolved (Tables 2.6, 2.7). Values of

Table 2.6 — ^{1}H-methyl NMR data for PMe$_2$Ph and its derivatives

	Me$_2$PhP	Me$_2$PhPO	Me$_2$PhPS	Me$_2$PhPH$^+$	Me$_2$PhPMe$^+$
δ(CH$_3$)	1.39	1.76	2.00	2.21	2.38
^{2}J(P–H)/Hz	1.7	13.0	13.2	15.2	13.2

Data from J. M. Jenkins and B. L. Shaw, *J. Chem. Soc.(A)*, (1966) 770.

Table 2.7 — ¹H-methyl NMR data for *cis*-PtX$_2$(PMe$_2$Ph)$_2$

X	Cl	Br	I	SCN
δ(CH$_3$)	1.77	1.87	1.95	1.77
^{2}J(P–H)/Hz	11.0	10.5	11.2	10.8
^{3}J(Pt–H)/Hz	35	36	37	34

Data from J. M. Jenkins and B. L. Shaw, *J. Chem. Soc.(A)*, (1966) 770.

10–15 Hz are typical for transition metal complexes. [This increase in ^{2}J(P–H) is attributed to the change in hybridization of the phosphorus atom from being close to p^3 plus a lone pair in the free ligand to sp^3 in the adduct, so that the *s*-character of the C–P bonds increases.] In platinum complexes, the spectrum is further complicated by the appearance of satellites due to ^{195}Pt with ^{3}J(Pt–H) about 35 Hz (see above, section 2.3.1.2). The spectrum of the ligand and *cis*-[PtCl$_2$(PMe$_2$Ph)$_2$] are illustrated in Fig. 2.7(a and b).

If two phosphine ligands are in mutually *trans* positions, the spectrum changes drastically, becoming apparently a triplet with a relatively small coupling constant. Thus, for *trans*-[PtI$_2$(PMe$_2$Ph)$_2$], illustrated in Fig. 2.7(c), the methyl signal appears at 2.20 δ with an apparent coupling constant of 3.6 Hz. Similar behaviour is seen in large numbers of bis-phosphine complexes, and has become an empirical rule:

For CH$_3$ [orC(CH$_3$)$_3$] groups in tertiary-phosphine complexes, doublet ¹H-spectra indicate mutually *cis* arrangements, and triplet spectra mutually *trans*.

This rule works well for square-planar and octahedral complexes of the 4*d* and 5*d* transition series. In other cases, quite complex spectra may be observed. Such curious behaviour is due to a phenomenon which has become known (inaccurately) as 'virtual' coupling. The name arises because, in the *trans* case, H$_3$CP(R$_2$)PMP(R$_2$)CH$_3$, the methyl protons seem to behave as if they were coupled equally to both phosphorus nuclei. This cannot be an accurate description since, relative to a given methyl group, the two phosphorus nuclei are not equivalent. It is also most unlikely that ^{2}J(P–H) would have the same value as ^{4}J(P–H). This is an example in which the simple first-order treatment is not applicable [cf. section 2.2.3.2]. Despite their apparent simplicity, the spectra are actually quite complicated.

The system with a single methyl group on each phosphine should be treated as A$_3$XMX′A$_3'$, in which the two sets of A nuclei (the protons) are equivalent chemically and symmetrically but not magnetically. The spectrum now consists of seven pairs of lines (2*n*+1, *n*=3), the positions and intensities of which depend on the relative magnitudes of the A–X (=A′–X′) and X–X′ coupling constants. In all cases, one pair of lines carries half the total intensity and has a separation of J(A–X)+J(A–X′) [=^{2}J(H–P)+^{4}J(H–P)]. Two other pairs are very weak and lie outside this main pair. The appearance of the spectrum is governed by the positions of the remaining lines. When J(X–X′) [^{2}J(P–P)] is small relative to

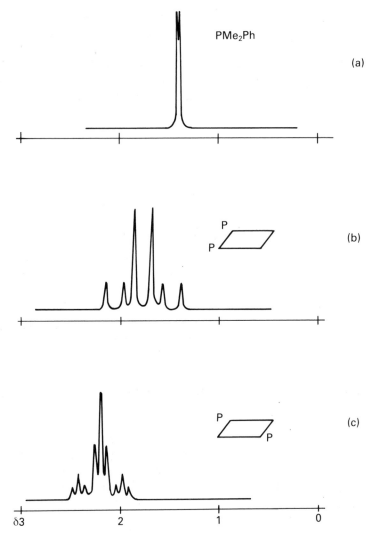

Fig. 2.7 — The methyl region of the ^{1}H spectra of (a) PMe$_2$Ph, (b) *cis*-PtCl$_2$(PMe$_2$Ph)$_2$, and (c) *trans*-PtI$_2$(PMe$_2$Ph)$_2$ (simulated from the reported parameters).

J(A–X) [^{2}J(H–C)], as in the *cis* complex, the remaining doublets fall close to the major doublet, and may not be resolved from it. The spectrum then approximates to a slightly broadened doublet. If J(X–X′) is large, as in the *trans* case, one doublet becomes very weak and has a large separation, and the remaining three doublets cluster together in the centre of the spectrum, which now approximates to a 1:2:1 triplet. Quite often the central region appears broader and less intense than would be expected for a simple triplet. Similar behaviour would be found for complexes of PMe$_2$Ph (*n*=6). The large value of ^{2}J(P–P) in the *trans* case is a manifestation of the *trans*-influence. [This description is based on the paper by R. K. Harris, *Can. J. Chem.*, **42** (1964) 2275.]

When three PMe$_2$Ph ligands are present in an octahedral complex, the most common stereochemistry is that in which they are coplanar, the *mer* configuration (*mer*=meridional). Two of the ligands are mutually *trans* and equivalent to each other, while the third has a different environment with a non-phosphine ligand *trans* to it. The ¹H spectrum then shows two sets of methyl signals: a doublet corresponding to the unique phosphine, and a triplet with double the intensity for the other two (Fig. 2.8).

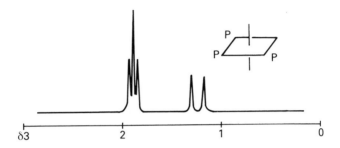

Fig. 2.8 — The methyl region of the ¹H spectrum of *mer*-IrCl$_3$(PMe$_2$Ph)$_3$ (simulated from the reported parameters).

In all the cases discussed so far, the two methyl groups of any given PMe$_2$Ph ligand have been equivalent. That is, each complex possesses a plane of symmetry (mirror plane) which passes through the phosphorus atom(s). If such a plane is not present, the two methyl groups become non-equivalent, and may have slightly different chemical shifts and coupling constants. (This remains true even if there is free rotation about the M–P bond since, except instantaneously, the two groups are in different environments.) In practice, the difference in parameters is quite small, and the two signals often overlap considerably (see Fig. 2.9).

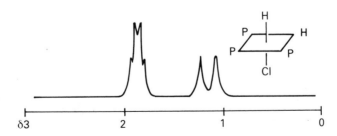

Fig. 2.9 — The methyl region of the ¹H spectrum of *mer*-IrH$_2$Cl(PMe$_2$Ph)$_3$, in which the two hydride ligands are mutually *cis*, making one pair of the methyl groups of each *trans* ligand non-equivalent (simulation).

2.3.3 Hydrides

Compounds containing hydrogen atoms bound directly to atoms other than carbon are conveniently subdivided into main group hydrides and transition-metal hydrides. These two classes of compound have rather different chemistry, and also give broadly different chemical-shift ranges.

2.3.3.1 Main group hydrides

Hydrogen atoms bonded directly to main group elements usually give chemical shifts in the conventional δ 0–10 range. The actual value of the shift depends principally on the effective electronegativity of the main group element in a similar way to that observed for organic compounds. That is, the shift depends on the basic electronegativity of the main group element plus the modifying effect of the other substituents (see Table 2.8). This means that the parent hydride often has a chemical shift rather

Table 2.8 — Chemical shifts and coupling constants for hydrogen bound to main-group elements

Pure gaseous hydrides (neat liquid)

	CH_4	NH_3	H_2O	HF
δ	$-0.24(-0.24)$	$-0.31(0.74)$	$0.31(4.89)$	$1.85(8.50)$
	SiH_4	PH_3	H_2S	HCl
	$2.75(1.91)$	$1.21(2.71)$	$-0.17(1.33)$	$-0.73(1.32)$
	GeH_4	AsH_3	H_2Se	HBr
	2.60	1.47	-2.31	-4.58
	SnH_4	SbH_3	H_2Te	HI
	3.84	1.38	-7.08	-13.49

Organometallics

	δ	$^1J(H-M)/Hz$
terminal B–H	-1 to $+2$	100–190 (^{11}B)
bridging B–H–B		<80 (^{11}B)
R_3SiH	3 to 5	150–350
R_2PH	4.5 to 5.5	180–220
R_2NH, RNH_2	0.5 to 5.0	
RSH	1.2 to 3.6	

Note: For many of these systems, the chemical shift depends markedly on the concentration of the solution and the nature of the solvent, as well as the electronegativity of the R-group. Increasing electronegativity of R gives an increase in chemical shift.

different from that of corresponding organo-element hydrides (e.g. PH_3 shows a doublet at δ 1.21 (in the gas phase), while primary and secondary organophosphines give doublets in the range δ 4.5–5.5).

Main group hydride chemical shifts are often very dependent on solvent, particularly when hydrogen bonding is possible. The hydrogen bonding concerned can be either with a polar solvent or by self-association, the latter being possible when the main group element has one or more lone pairs. Hydrogen bonding usually results in a shift to higher frequency, which can be several ppm.

When a hydride containing a lone pair (e.g. a secondary phosphine) co-ordinates to a transition metal, the hydride resonance moves to higher frequency (e.g. 2–3 ppm for co-ordinated secondary phosphines).

Hydrogen atoms which form electron-deficient bridges, as in boron and aluminium hydrides, usually give signals with lower chemical shifts (and coupling constants) than would otherwise be expected.

2.3.3.2 Transition-metal hydrides

In many complexes, especially those involved in catalytic processes, one or more hydride ligands may be present. It is almost impossible to detect them by conventional elemental analysis, especially if organic ligands are also present, and they may even escape detection in the infra-red spectrum. In the NMR spectrum they are readily recognized by their characteristic chemical shifts, which are almost always to low frequency (high field) of TMS (i.e. $\delta < 0$). The most common range is $\delta - 10$ to -30, but signals have been observed between $\delta + 3$ and -60. These apparently anomalous shifts for transition-metal systems indicate that the hydrogen nucleus is experiencing considerable shielding, which arises in two ways. First, the ligand may actually be hydridic rather than protonic; in many cases it reacts with acids and not with bases. This suggests that there is relatively high electron density associated with the hydrogen ligand, considerably more than in the covalent main group hydrides, which usually react more readily with bases than with acids. The ligand is therefore formulated as a co-ordinated hydride, H^-, with a formal $1s^2$ configuration. Second, the metal atom usually has many d electrons in its valence shell, some of which have π symmetry with respect to the M–H bond; they tend to 'engulf' the hydride, contributing further to the shielding.

In addition to ensuring that the correct chemical shift region is being searched, one further precaution is very necessary in attempting to establish the presence of hydride ligands in the presence of organic ligands: a good signal : noise ratio must be obtained. This is because the relative intensity of the signal due to the hydride is frequently very small, especially when it is spread out by coupling with other nuclei. This effect is often compounded by low solubility. A further complication is that some hydrides react with chlorinated solvents (e.g. $CDCl_3$) to give the corresponding metal chloro-complex and the reduced solvent (e.g. $CHDCl_2$).

Much structural information may be gained from the coupling patterns exhibited by hydride resonances. Owing to the unique region in which the signals occur, they do not usually overlap with those of the other ligands, so that complete patterns can be readily discerned. Significant coupling may occur with nuclei of other ligands, especially phosphines, or with the metal nucleus. Coupling to other hydride ligands is normally quite small (1–5 Hz), and may well not be resolved; the protons of organic ligands are too far removed for coupling to be observed.

2.3.3.3 Coupling to phosphine ligands

Since ^{31}P has $I = 1/2$ and 100% abundance, the splitting of a hydride signal in a tertiary–phosphine complex is precisely analogous to that of a proton in an organic group which has other protons in neighbouring positions. Thus, the presence of a single phosphine ligand will split the hydride signal into a doublet. The magnitude (and sign) of the coupling constant depends on the P–M–H bond angle. When the phosphine is in a positon *cis* to the hydride (90°), the coupling constant $^2J(P-H)$ is typically 10–20 Hz, but for the *trans* case (180°) it is considerably greater, 100–200Hz. Two equivalent phosphines (necessarily in *cis* positions relative to the hydride) will give a symmetrical $1:2:1$ triplet, three give a $1:3:3:1$ quartet and four a $1:4:6:4:1$ quintet. All follow the familiar $2n+1$ rule and the intensities are in the ratios of the appropriate binomial coefficients (see section 2.2.2). The last case has, of course, no counterpart in organic chemistry, but could occur in an octahedral metal complex.

It is occasionally necessary to examine the relative intensities of the coupling pattern with considerable care, especially if the spectrum is rather noisy. This is because the whole pattern may result from a single hydride ligand, so that the integration should correspond to the value for one hydrogen atom. As the number of splittings increases, the individual signals become weaker and, unless the signal: noise ratio is increased, the outermost lines, which are the weakest, may merge into the baseline noise. It is then necessary to take account of the maximum possible number of equivalent couplings which may occur, and to estimate the relative intensities of the visible lines as accurately as possible and check that they are as expected. If the outer lines of, for instance, a quintet escape detection, their presence should be indicated by the fact that the remaining three lines have intensities in the ratio $2:3:2$; this ratio does not, of course, correspond to that of symmetrical triplet, which should be $1:2:1$.

In many cases, the phosphine ligands will not all be equivalent. For instance, two phosphines in *cis* positions in an octahedral complex may each be *trans* to a different ligand. The two coupling constants $^2J(P-H)$ to a hydride *cis* to both of them are then likely to be different. The expected coupling pattern is therefore a doublet of doublets, with all lines of the same intensity (see Fig. 2.2). Depending on the line widths and the difference between the two J values, the two central lines may or may not be resolved; in the latter case, the spectrum will approximate to a triplet.

Three phosphine ligands can only be equivalent in tetrahedral, trigonal-bipyramidal or fluxional octahedral geometry. In this case, of course, the spectrum is the symmetrical quartet described above. More usually, two of the phosphines are equivalent, giving a triplet pattern which is doubled by coupling with the third phosphine. This last may be either *cis* or *trans* to the hydride, and the observed patterns reflect this difference (Fig. 2.10). It is useful to be able to recognize these patterns from verbal descriptions when the spectra themselves are not available, e.g. when reading the literature. It is usual to refer to the components of the coupling scheme in order of decreasing magnitude of the coupling constants. Thus, the two spectra of Fig. 2.10 would be referred to as a triplet of doublets or a doublet of triplets according to whether the triplet or the doublet splitting is the larger. The nomenclature is obvious when applied to well-resolved patterns. In drawing the J-trees, greatest clarity is achieved by taking the coupling constants in order of decreasing magnitude. The final pattern is, of course, independent of the order chosen (cf. Fig. 2.3).

Coupling to four phosphorus atoms may occur in an octahedral tetrakis–phosphine complex or in a fluxional five-coordinate system. The phosphorus atoms are usually equivalent, the major exception being in some complexes involving two bidentate phosphine ligands.

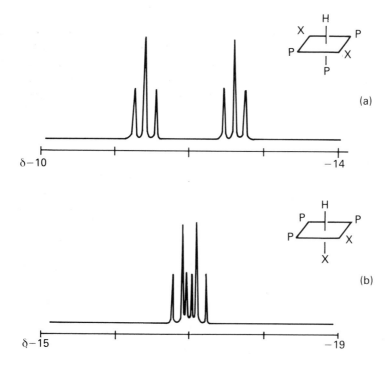

Fig. 2.10 — The hydride regions of the ¹H spectra of two metal hydrides showing coupling to two equivalent and one unique ³¹P nucleus. The parameters used are typical of the geometries shown: (a) δ−12, ²J(H–P)=14, 14, and 120 Hz; (b) δ−17, ¹J(H–P)=13, 13, and 19 Hz. Note that in spectrum (b) the two triplets overlap (simulations).

2.3.3.4 More than one hydride
When two or more hydride ligands are present, they are often equivalent, either by symmetry of by fluxionality, and the spectrum simply gains in intensity while retaining the expected shape. Sometimes, however, the hydrides may be chemically and symmetrically equivalent, but magnetically non-equivalent, as, for example in the (non-fluxional) square-planar complex **2.XIII**.

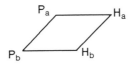

2.XIII

Although the coupling constants are equal in pairs,

$$^2J(P_a-H_a)=\,^2J(P_b-H_b); \qquad ^2J(P_b-H_a)=\,^2J(P_a-H_b)$$

each proton couples differently to the two phosphorus nuclei,

$$^2J(P_a-H_a)\neq\,^2J(P_b-H_a); \qquad ^2J(P_a-H_b)\neq\,^2J(P_b-H_b)$$

Because of this latter fact, the protons (and the phosphorus nuclei) are magnetically non-equivalent. The spectrum is no longer first-order and may be quite complex (see Prob. 2.2).

When the hydride ligands are non-equivalent by symmetry and have appreciably different chemical shifts, each will exhibit its own first-order coupling pattern. The problem is now one of recognizing these patterns, which will be of the types described above. If the patterns overlap, there may be some disentangling to be done. Recognition of the individual patterns is aided by remembering that the line spacings, as well as the intensities, are regular.

Coupling between the hydride ligands themselves is always small, seldom more than a few hertz, and is often not resolved. When it is resolved, it should be relatively easy to recognize, since every peak of the whole pattern is split in exactly the same way. An example is shown in Fig. 2.11.

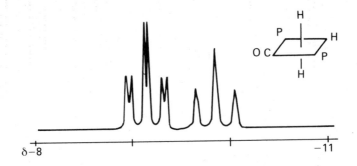

Fig. 2.11 — The hydride region of $IrH_3(CO)(PPh_3)_2$. Simulated from the reported parameters: $\delta-9.15$, $^2J(H-P)$ 17.0 Hz, $^2J(H-H)$ 4.5 Hz; $\delta-19.85$, $J(H-P)$ 20.0 Hz, $^2J(H-H)$ 4.5 Hz).

2.3.3.5   Coupling to the metal

Some transition-metal nuclei have non-zero spins, which give rise to additional coupling. Two common cases with spins of one-half will be described: rhodium and platinum. Extrapolation to other cases should then be simple.

The effect of an additional spin of one-half is simply to introduce an additional splitting of all lines by $^1J(M-H)$. Since the hydride is directly bonded to the metal, the coupling should be strong, and its magnitude will depend on the size of the nuclear magnetic moment. For ^{103}Rh, the moment is quite small and $^1J(Rh-H)$ is usually only about 15 Hz. It is thus comparable with many of the other coupling constants which may be involved, and the spectra may become rather complicated. Some typical examples are shown in Fig. 2.12. $^1J(Rh-H)$ is similar in magnitude to

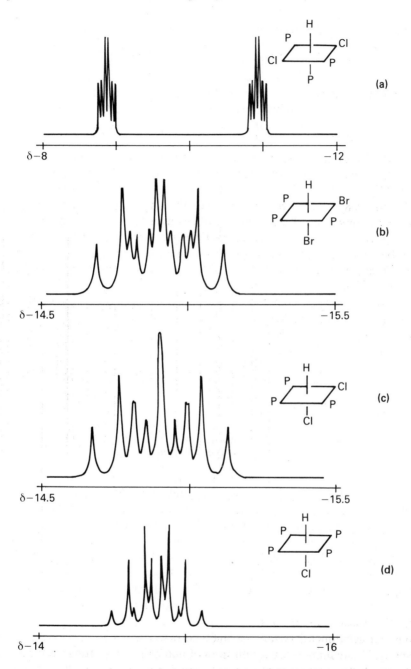

Fig. 2.12 — The hydride region for various rhodium(III) complexes. Simulated from the reported parameters for complexes of PPh₂Me for (a)–(c) and Ph₂P(CH₂)₂PPh₂ for (d): (a) $\delta-9.9$, doublet [^{2}J(H–P) 206 Hz] of triplets [^{2}J(H–P) 9 Hz] of doublets [^{1}J(H–Rh) 4 Hz]; (b) $\delta-14.9$, doublet [^{1}J(H–Rh) 14 Hz] of doublets [^{2}J(H–P) 11.5 Hz] of triplets [^{2}J(H–P) 9 Hz]; (c) $\delta-14.9$, doublet [^{1}J(H–Rh) 14.5 Hz] of doublets [^{2}J(H–P) 13.5 Hz] of triplets [^{2}J(H–P) 9 Hz]; (d) $\delta-14.8$, doublet [^{1}J(H–Rh), 15.5 Hz] of quintets [^{2}J(H–P) 11.5 Hz]. Note the overlapping multiplets in (b) and (c).

$^2J(P_{cis}\text{–H})$, and it is often difficult to assign the coupling constants without recourse to the ^{31}P spectrum.

With platinum, the nuclear magnetic moment is large, and the bonds are more covalent than for rhodium. The coupling constants are usually much larger than those to other nuclei, so that a simple doubling of the whole pattern is seen. There is, however, a different complication in this case. While natural rhodium is 100% ^{103}Rh, natural platinum exists as a mixture of several isotopes, only one of which as a spin: ^{195}Pt, 33%, $I=1/2$. The spectra of platinum complexes therefore appear as those of a mixture: two-thirds of the molecules display only the simple coupling pattern due to the ligands, while the remaining one-third show additional doubling due to coupling to ^{195}Pt. Owing to the large magnetic moment, $^1J(Pt\text{–H})$ is large, typically in the range 700–1400 Hz. The additional signals thus appear as satellites, well separated from the main signal, with one quarter of its intensity; both sets, of course, have the same chemical shift. For instance, in a 100 MHz spectrum with $^1J(Pt\text{–H})=1100$ Hz, the satellites would appear at ±5.5 ppm from the central signal (Fig. 2.13). In some

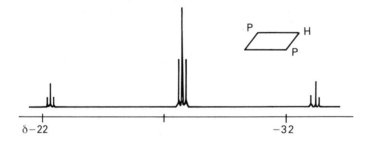

Fig. 2.13 — The hydride region of the spectrum of *trans*-PtH(NCS)(PEt$_3$)$_2$ (simulated from the parameters in Table 2.9).

cases. $^1J(Pt\text{–H})$ may be larger than this, and it is quite possible that one of the satellites will not appear on a conventional 10 ppm-wide scan. (In FT spectra, one of the satellites may be folded in, and appear as an inverted peak.) Each satellite has the same shape as the main signal, since the coupling patterns to the other nuclei are still present. Another way of establishing that the outlying signals are satellites, and are not due, for instance, to an impurity, would be to remeasure the spectrum with a different spectrometer frequency. The chemical-shift value at which the satellite signals appear will change, but the average of the two (the true chemical shift) will remain identical to that of the main signal. The frequency-separation between the satellites will be unchanged, being $^1J(Pt\text{–H})$ (see Fig. 2.14).

When the metal atom has a spin greater than 1/2, the splitting pattern consists of $2I+1$ peaks, but all have the same intensity. For instance, a hydride bound to ^{51}V, with spin of 7/2, gives an eight-line pattern as shown in Fig. 2.15 for $[HV(CO)_5]^{2-}$.

2.3.3.6 *Effect of* trans *ligand*
The chemical shift of a hydride ligand depends on many factors, e.g. the co-ordination number of the complex, the transition series to which the metal belongs,

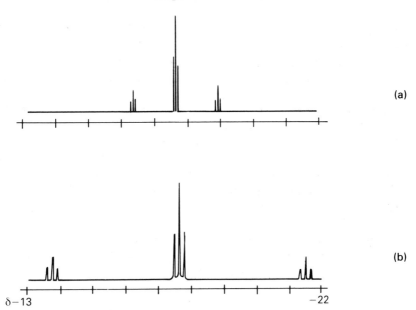

Fig. 2.14 — The hydride region of the spectrum of *trans*-PtH(CN)(PEt₃)₂, at spectrometer frequencies of (a) 300 MHz and (b) 100 MHz. (Simulated from the data in Table 2.9.)

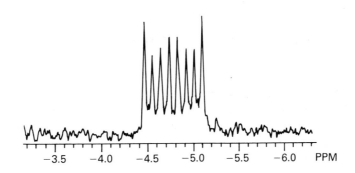

Fig. 2.15 — The ¹H spectrum of Na₂[HV(CO)₅] (reproduced with permission from G. P. Warnock and J. E. Ellis, *J. Amer. Chem. Soc.*, **106** (1984) 5017).

and the identity of the other ligands. Within a series of closely related complexes, the last effect is often systematic, and can be used to assist the diagnosis of structures.

In such a series of complexes, the chemical shift (and, to some extent, the coupling constants) of a hydride ligand reflect principally the identity of the ligand in the *trans* position. This appears to be a manifestation of the *trans*-influence which is also apparent in the M–H stretching frequencies. Thus for the series of square-planar

Table 2.9 — ^{1}H NMR data for *trans*-[PtHX(PEt$_3$)$_2$] (in benzene)

X=	ONO$_2$	NO$_2$	NCO	NCS	Cl	Br	SCN	I	CN
δ(H)	−23.6	−19.4	−27.7	−27.6	−26.8	−25.5	−23.0	−22.7	−17.6
^{1}J(Pt–H)/Hz	1322	1003	1080	1086	1275	1346	1233	1369	778
^{2}J(P–H)/Hz	15.6	16.7		14.0	14.5	13.8	14.5	13.3	15.6
ν(Pt–H)/cm^{-1}	2242	2150	2234	2195	2183	2178	2160	2156	2041

Data from J. Chatt and B. L. Shaw, *J. Chem. Soc.*, (1962) 5075.

platinum(II) complexes *trans*-[PtHX(PEt$_3$)$_2$], the hydride resonance moves to higher frequency as the *trans*-influence of X increases (Table 2.9). Similar trends are seen in the corresponding arsine complexes, and in six-coordinate complexes, for instance those of rhodium(III), iridium(III) or ruthenium(II) (Table 2.10).

Table 2.10 — Typical ranges of ^{1}H NMR data for six-coordinate hydride complexes of Group VIII metals

trans-ligand	Halide	Phosphine/Arsine	Silyl/Germyl	CO
δ(H)	-15 to -22	-10 to -15	-8 to -10	*ca* -9
^{2}J(P$_{cis}$–H)/Hz	12 to 20	19 to 20	15 to 16	*ca* 17

In all cases, the coupling constants to phosphine ligands show little systematic variation, and serve only to indicate whether the ligand is *cis* or *trans* to the hydride.

2.3.3.7 Dihydrogen complexes

In recent years it has been recognized that some transition-metal hydrides may involve a sideways bonded (η^2) dihydrogen molecule rather than two σ-bonded hydride ligands. Examples are [FeH(H$_2$)(dppe)$_2$]BF$_4$ and [W(H$_2$)(CO)$_3$(PR$_3$)$_2$]. The dihydrogen ligand is not easy to recognize, since its ^{1}H NMR signals occur in the normal hydride region, albeit at the high-frequency end: $\delta-3$ to -15. In several cases, conventional ('classical') hydride ligands are also present, and rapid exchange occurs at room temperature between the two types, or there may be equilibrium between the dihydrogen and dihydride forms: the two complexes mentioned above display such behaviour. It is then necessary to cool the sample in order to resolve the signal for the dihydrogen ligand. In the absence of exchange effects, the signals are usually very broad, with line half widths of 3–300 Hz: The large line width is the result of rapid dipole–dipole relaxation of one hydrogen atom by the other, which is extremely efficient when the two are directly bonded. It has been recommended that the presence of a dihydrogen ligand is best defined by a measurement of the relaxation time (T_1): for normal hydride ligands, T_1 is usually several seconds or more, whereas bound dihydrogen has (to date) been observed to give values of 4–100 ms. An alternative diagnosis, can sometimes be made by preparing the HD analogue, when ^{1}J(H–D) may be seen in the spectrum [^{2}D, $I=1$]. Values of 22–32 Hz have been recorded. However, if the complex contains both dihydrogen and conventional hydrides, equilibration between them will almost certainly result in a mixture of complexes containing all permutations of H and D.

2.4 ISOTOPES OTHER THAN ^{1}H

Complementary information can usually be obtained by probing other nuclei in the molecule. A wide range of isotopes is now available including those of several of the metals. Whatever the isotope probed, the principles of interpretation are always the same: one looks for the number of discrete sets of signals and the patterns of coupling to other active nuclei. The commonest problem for the inexperienced is remembering that the viewpoint has changed — the system is now being seen by an observer standing on the new nucleus and no longer by a 'hydrogen observer'. Thus, for instance, the only observable coupling constants in a ^{31}P spectrum are those involving the ^{31}P nucleus.

The most popular isotopes are ^{31}P, in complexes of tertiary phosphines, ^{13}C (organometallics and carbonyls), and ^{29}Si (in silicates and zeolites), but many others are used, such as ^{14}N (amines, etc.), ^{59}Co, ^{103}Rh, and ^{195}Pt, and all of these are discussed below. Other examples are given in the problems.

In order to simplify the spectra, the coupling to protons is often removed, by a technique known as **broad-band decoupling**. An additional band of frequencies is applied at the same time as the exciting frequency for the nuclei under investigation. The extra frequencies cover the range in which most ^{1}H nuclei resonate, and this effectively excites and de-excites the protons so rapidly that their coupling effects are averaged to zero. To indicate that this has been done, the spectra are usually designated ^{31}P–$\{^{1}$H$\}$, etc. The technique is particularly useful when organic groups are bound directly to the atom being probed. If necessary, the decoupling can be made selective, so that the coupling of particular groups of protons can be removed or retained. This is valuable, for instance, for metal hydride complexes containing phosphines, where one might wish to remove the coupling by organic groups but retain that due to the hydride ligand.

Other nuclei can also be decoupled if necessary, which can often help the assignment of signals for complicated molecules; an example of the decoupling of ^{103}Rh in a ^{13}C spectrum is given below (section 2.4.2). The removal of coupling also means that a satisfactory spectrum can be accumulated more rapidly, since the intensities of the individual signals are split between fewer lines. However, there are sometimes penalties to be paid for these advantages.

First, much of the energy of the decoupling radiation remains in the sample, and there may be a considerable temperature increase. Second, the rapid relaxation of the irradiated nuclei can also affect the nuclei under observation, changing their energy distribution. The immediate result of this is that the intensities of the signals change, and integrations are no longer reliable. The size of this effect, the **nuclear Overhauser effect** (NOE), depends on the magnetogyric ratios of the nuclei under observation (A) and those being irradiated (X), so that the relative intensity of a signal becomes $1+\gamma_X/2\gamma_A$. For some nuclei, γ is negative (e.g. ^{15}N, ^{29}Si), so that it is possible for a signal to have negative, or even zero intensity. The full effect of NOE is often not seen and, in any case, it depends on the proximity of the two nuclei. The net result is that relative intensities are no longer a safe guide to the numbers of nuclei in different sites. If it appears that some signals have anomalous intensities or (for the nuclei mentioned above) to be missing altogether, it is wise to obtain the undecoupled spectrum for comparison.

In contemplating the use of one of the less common isotopes, several factors have to be taken into consideration. First, is the frequency within the range of any spectrometer available? Second, is a signal likely to be easily detectable? Third, is the spin likely to cause complications? Relevant data are provided in Appendix 1. The first question is almost self-evident, and can only be answered locally. The second is best estimated from the **relative receptivity** for the nucleus in question. This quantity is the sensitivity to detection relative to the same number of protons, and includes a weighting for the natural abundance of the isotope. Ideally, this factor should be further weighted (reduced) by the relaxation time (line width) and various instrumental factors. Very small values of the receptivity can be tolerated (e.g. 10^{-4} or less) provided adequate accumulation time is available. If proton decoupling is to be used, the sign and magnitude of the magnetogyric ratio of the isotope, γ, become important, as shown in the preceding paragraph. Finally, as discussed in section 2.2, line widths are often broadened for isotopes with spin greater than 1/2 through quadrupole coupling effects. When there are two quadrupolar isotopes of the same element, the sharper line will be obtained by the isotope with the smaller value of $(2I+3)(eQ)^2/[I^2(2I-1)]$, where eQ is the nuclear quadrupole moment (see Chapter 3).

A survey of some of the more useful nuclei is given below. For more information on these and other isotopes, the books edited by Mason and by Laszlo and the references therein should be consulted (see Bibliography).

2.4.1 Phosphorus-31
Phosphorus-31 is probably the next most readily available nucleus after ^{1}H and possibly ^{13}C. This is because the isotope has 100% abundance, resonates in a readily attainable range of frequencies, and has relatively high receptivity. The nuclear magnetic moment is about 40% lower than that of ^{1}H, and a correspondingly lower frequency is used for any given magnetic field. An operating frequency of about 30 MHz is common, which corresponds to about 90 MHz for ^{1}H (Table 2.1). However, much better resolution is usually achieved than for ^{1}H because the chemical shifts cover a very wide range (more than 500 ppm, Fig. 2.16).

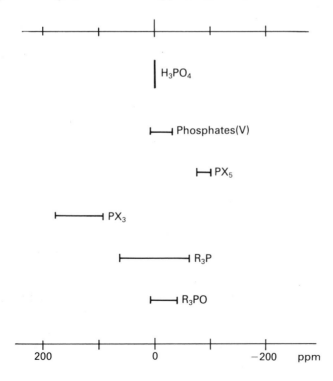

Fig. 2.16 — Typical ranges for ^{31}P chemical shifts. (X=F, Cl, Br, I, or OR; R=alkyl or aryl.)

^{31}P is one of the nuclei for which care has to be taken when consulting the literature, owing to the redefinition of the sign of the chemical-shift scale. Chemical shifts are now quoted as positive when they are to high frequency (low field) of the standard; the older literature used the opposite convention. Care should therefore be taken with values given in any work earlier than about 1975. For transition-metal complexes, a simple check is to look for the chemical shift quoted for the ligand itself: values for some common ligands are given in Table 2.11.

Chemical shifts are usually referred to 85% phosphoric acid as standard. This is not normally a desirable material to add directly to the solution under investigation, but can be inserted in a sealed capillary. It is also often necessary to work at temperatures below the freezing point of phosphoric acid, e.g. when there are ligand-exchange effects. In these cases some other suitable standard must be found. In co-ordination chemistry, the ligand itself is often satisfactory, either in the solution or in a capillary; its chemical shift relative to H_3PO_4 can be determined separately. Note that if the ligand is added directly to the solution it may exchange with the co-ordinated ligands, even at low temperatures.

Much early work with ^{31}P was concerned with identification of the various polyphosphates present in aqueous solution, examples of which are shown in Fig. 2.17. The technique is also very useful in the characterization of P–N, P–B, P–Si and other systems which may involve oligomer formation. In recent years, however, work has concentrated on tertiary-phosphine complexes of transition metals, and the principal discussion here will be focussed on organophosphines and their derivatives.

The spectra of compounds containing organic groups are usually obtained with proton decoupling, which gives considerable simplification. Thus, for instance, PPh_3 gives only a single signal instead of the very complicated pattern due to the coupling by six *ortho*, six *meta*, and three *para* protons. In a $^{31}P-\{^1H\}$ spectrum, each distinct type of phosphorus atom gives a single signal showing only the coupling to the other ^{31}P nuclei (unless other non-hydrogen nuclei with spins are present, such as metal atoms). As discussed above, decoupling may affect the relative intensities of the signals. Samples containing only metal-bound phosphines often give fairly reliable integrations, but these cannot be related to that for any free ligand which may be present. If accurate integrations are required, the spectrum should be obtained without decoupling; with FT spectrometers, it is also advisable to increase the pulse delay, to check that the relative intensities do not change (metal complexes usually give shorter relaxation times then the free ligands).

The spectrum of an organophosphine ligand is often recorded in order to check on its purity, since the corresponding phosphine oxides have distinctly different chemical shifts. Comparison of the shift with that of the complex under investigation is also useful. The chemical shifts of some tertiary organophosphines and their oxides are given in Table 2.11. With the majority of common ligands, the oxide gives a signal to high frequency of the phosphine.

The increase in chemical shift which occurs on oxidation of a tertiary phosphine is

Table 2.11 — Chemical shifts of tertiary phosphines and their oxides (to high frequency of H_3PO_4)

R_3	Me_3	Et_3	$^nPr_3$	$^nBu_3$	Ph_3	Ph_2Me	Ph_2Et	$PhMe_2$	$PhEt_2$
$\delta(R_3P)$	−62	−20	−33	−33	−7	−38.5	−12	−46	−18
$\delta(R_3PO)$	36	48		43	29				42

Note: These values are somewhat solvent-dependent, and reported values may differ by *ca* ±2 ppm.

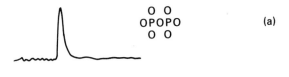

Fig. 2.17 — The ^{31}P spectra of (a) $P_2O_7^{4-}$, (b) $P_3O_{10}^{5-}$, and (c) $P_4O_{13}^{6-}$.

also seen when it becomes co-ordinated to a metal or to an organic group or to a proton. For example, the tri- and tetra-phenylphosphonium cations, Ph_3PH^+ and Ph_4P^+, give signals at -1.0 and $+21$ ppm. In simplistic terms, this represents donation of the lone pair to an oxygen atom, or to a cation (H^+, M^{n+}, or R^+), resulting in a deshielding of the phosphorus nucleus. However, this is not an invariable rule, since some transition metal complexes show the opposite behaviour (e.g. iridium(III), see below). For tri-organoxyphosphines, $(RO)_3P$, and the phosphorus trihalides the shift is also in the opposite direction, e.g. the shifts for $(PhO)_3P$ and $(PhO)_3PO$ are $+127$, and -1.0 ppm, and for Cl_3P and Cl_3PO, $+219$ and $+2$ ppm. These ligands often show a decrease in chemical shift on co-ordination to a metal also. Much depends on the hybrid character of the lone pair and the bond pairs and the way in which they change on involvement of the lone pair in a bond.

For transition-metal complexes, it is useful to define a quantity known as the **co-ordination chemical shift** [$ccs = \delta(\text{complex}) - \delta(\text{ligand})$], which often shows systematic variations. For instance, in series of complexes in which the metal atom has a d^{10} electron configuration, e.g. Pt^0, Au^I, the ccs decreases systematically as the co-ordination number increases (Table 2.12). As the number of ligands bound to any one metal increases, so the shift becomes closer to that of the free ligand. When analogous complexes of a series of tertiary phosphines are compared, there is often a linear relationship between the chemical shift of the free ligand and the co-ordination chemical shift. Data for some complexes are shown in Fig. 2.18, which indicates good

Table 2.12 — Co-ordination chemical shifts for some d^{10} complexes

M	Ligand	δML_2	δML_3	δML_4
Au$^+$	PEt$_3$	62.4	57.4	13.4
	PMePh$_2$	54.5	42.3	22.8
	PMe$_2$Ph	63.1		22.8
	PBu$_3$	73.5	70.3	
	PCy$_3$	54.3		
Ag$^+$	Ptol$_3$	22.0	19.5	14.5
Pt^0	PEt$_3$		64.0	4.6
	PBu$_3$		65.6	7.7
	PCy$_3$	53.6	45.0	
Pd^0	PEt$_3$		9.6	−1.5
	PBu$_3$		−1.4	−7.9
	PCy$_3$	38.7	25.9	

Data from R. V. Parish, O. Parry and C. A. McAuliffe, *J. Chem. Soc., Dalton Trans.* (1981) 2098, and references therein.

correlations provided the ligands are neither very bulky nor highly asymmetric. Similar correlations have been found for many other types of complex.

The *ccs* values shown in Table 2.12 and Fig. 2.18 are all positive, i.e. co-ordination has increased the chemical shift of the phosphorus atom. In some cases the opposite is observed, e.g. for *mer*-[IrI$_3$(PMePh$_2$)$_3$] the two types of phosphine show *ccs* values of −17.2 and −24.8 ppm (*trans* to I$^-$ and to PMePh$_2$ respectively, see structure **2.XIV**), and in the corresponding chloro-complex the values are −2.6 and +6.5 ppm. Similarly, the *ccs* for many platinum complexes are negative, especially when the substituents are bound to phosphorus through oxygen or nitrogen atoms.

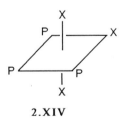

2.XIV

In complexes containing non-equivalent phosphine ligands, the effects of P–P coupling can be seen. For instance, in *mer*-[IrCl$_3$(PEt$_3$)$_3$] two ligands are equivalent but different from the third. The ^{31}P–{^{1}H} spectrum therefore shows a doublet for the two equivalent ligands (split by the third) and a triplet of half the intensity for the unique ligand (coupling to the other two). Both sets of signals are characterized by the same coupling constant, ^{2}J(P–P) (16.5 Hz, see Fig. 2.19(a)). The spectra of the

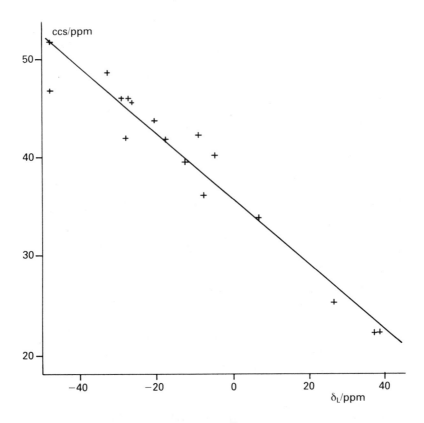

Fig. 2.18 — Coordination chemical shift versus ^{31}P chemical shift of the ligand for *trans*-RhCl(CO)(L)$_2$, where L is a tertiary phosphine. (Redrawn from the data of B. E. Mann, C. Masters, and B. L. Shaw, *J. Chem. Soc. (A)*, (1971) 1104.)

corresponding rhodium(III) complex is similar but each signal is further doubled by coupling to ^{103}Rh (100%, $I=1/2$). However, since the ligands are not equivalent, two different coupling constants are involved, that to the unique ligand being the larger [$^1J(Rh–P)=112$ and 84 Hz] (see Fig. 2.19(b)).

The difference in magnitude of the two Rh–P coupling constants is due to the *trans*-influence, since the larger value of $^1J(P–Rh)$ is associated with the phosphine ligand atom *trans* to a chloride ligand rather than to another phosphine. Coupling constants are (in part) a measure of the electron density in the bond system (strictly, the s-electron densities). A ligand which exerts a strong *trans*-influence weakens the bond opposite to itself and reduces the electron density in that bond. Coupling which involves the weakened bond is therefore reduced. This effect is nicely seen in the *cis* and *trans* isomers of platinum complexes, $PtX_2(PR_3)_2$. As the data of Table 2.13 show, the phosphorus–platinum coupling is weaker for a phosphine which is *trans* to another phosphine than for one *trans* to an anionic ligand with a low *trans*-influence. (Remember that, as discussed in sections 2.2.3.1 and 2.3.3.5 above, the coupling in platinum complexes is shown by the satellites, and J-values are measured between

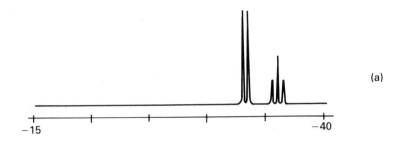

(a)

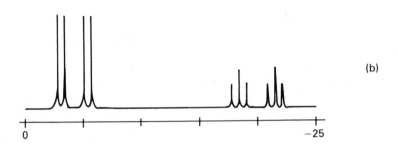

(b)

Fig. 2.19 — ³¹P-{¹H} spectra of (a) *mer*-IrCl₃(PEt₃)₃ and (b) *mer*-RhCl₃(PEt₃)₃. (Simulated from the reported data — note that, when spectrometer frequency is relatively low, the intensities may be less regular than the idealized doublets and triplets shown, owing to second-order effects.)

Table 2.13 — ³¹P nmr data for *cis* and *trans* PtX₂(PBu₃)₂

	cis		*trans*	
X	δ	$^1J(P{-}^{195}Pt)/Hz$	δ	$^1J(P{-}^{195}Pt)/Hz$
Cl	−1.4	3508	−4.9	2380
Br	−0.8	3479	*ca* 0	2334
I	+1.1	3372	+7.9	2200

A. Pidcock, R. E. Richards and L. M. Venanzi, *J. Chem. Soc., A* (1966) 1707.

the two satellite peaks, and not from the major, central signal.) Thus, although the chemical shifts are of little value in determining the stereochemistry, the coupling constants are quite distinctive. In general, for platinum(II) dihalide complexes, $^1J(P{-}Pt)$ is 2000–2500 Hz for a phosphine *trans* to another phosphine, and 3000–3500 Hz for a phosphine *trans* to a halide. Values for other platinum(II) systems may lie outside these ranges, but the trends are always the same. For instance, *cis*-

$PtCl(CH_3)(PEt_3)_2$ has non-equivalent phosphine ligands, which show chemical shifts of 8.7 and 14.6 ppm, and coupling constants of 4179 and 1719 Hz for the groups *trans* to the chloride and to the methyl, respectively. The *trans* isomer shows a single signal at 16.2 ppm, with a coupling constant of 2821 Hz. The *trans*-influence series X<P<C<H (X=Cl, Br, or I) applies for most metal complexes.

Precisely similar trends are found for platinum(IV) complexes, or for complexes of other metals (e.g. rhodium, see Table 2.14). It is, of course, necessary to make

Table 2.14 — Typical ranges for metal–phosphorus coupling constants, $^1J(P-M)/MHz$, for various stereochemistries

	cis	trans
Octahedral rhodium(III)	100–120	80–90
Square-planar rhodium(I)	160–190	120–140
Octahedral platinum(IV)	2000–2500	1400–1800
Square-planar platinum(II)	3000–3500	2000–2500

comparisons only between related complexes. For instance, $^1J(P-Pt)$ for the platinum(IV) complexes *cis*- and *trans*-$PtCl_4(PBu_3)_2$ are 2065 and 1474 Hz, respectively. The decrease in phosphorus–metal coupling with increase in oxidation state of the metal appears to be general and derives from the associated changes in co-ordination number and hybridization. In the square-planar/octahedral comparison, the *s*-character of the metal–phosphorus bonds decreases from 1/4 to 1/6, and the coupling constants are approximately in the same ratio. This implies that there is relatively little change in covalency (*s*-character or the bonds) from one oxidation state to the other. However, the same comparison cannot be made between square-planar platinum(II) and tetrahedral platinum(0), in both of which the *s*-character is 25%. For the tetrahedral complexes, $^1J(P-Pt)$ is 3600–3900 Hz; the value may be even higher if dissociation to three- or two-co-ordinate platinum(0) occurs.

All the above trends are also found for chelating ligands containing phosphine groups. In addition, there may be a systematic effect on the co-ordination chemical shift: diphosphine ligands in five-membered chelate rings show high-field shifts of some 25 ppm relative to the *ccs* of related monodentate ligands in complexes with similar geometry, while those in rings of other sizes (four-, six- or seven-membered) show a range of shifts to lower frequency.

Tridentate phosphine ligands, such as $PhP(CH_2CH_2CH_2PPh_2)_2$, are particularly easy to study, since the central and terminal phosphine groups must always be inequivalent, and can readily be recognized from the difference in intensities and coupling patterns (the phosphorus nuclei are coupled to each other with J values of several tens of Hz).

Confusing effects are sometimes found if the sample contains traces of the free ligand. This is particularly true of complexes which are labile to substitution, especially those with d^{10} configurations. In these cases, rapid exchange between the co-ordinated ligand and the free ligand may occur at room temperature, with the result that only an average signal is seen. Similar exchanges can occur in solutions containing more than one complex. It is then necessary to lower the temperature of measurement, often very considerably. For instance the data for the gold complexes given in Table 2.11 were obtained at 180 K; this is about the lowest temperature which can be readily obtained, and restricts the solvents which can be used (dichloromethane was used in this case). For the gold system, exchange is still rapid even at this temperature when the ligand is PPh_3. It is possible, in principle at least, to identify the complexes being formed by plotting the ambient temperature (averaged) chemical shift against the metal–ligand ratio. Provided that at most two complexes are present for any particular composition, a series of straight lines should be obtained with breaks indicating the position of stable complexes.

2.4.2   Carbon-13

Carbon-13 has a wide range of chemical shifts and narrow lines (when ¹H decoupling is employed, as is usual), factors which make for good resolution. Frequently a structure may be confirmed (or otherwise) simply by counting the number of separate signals, which should agree with the number of distinct carbon sites.

There are several minor problems associated with measurements on ¹³C in organometallic compounds.

(a) The isotope has very low abundance (1.1%), which means that long spectrum-accumulation times may be required, especially if the sample has low solubility. Signal intensity is improved by proton decoupling, and the nuclear Overhauser effect (see section 2.4) can further enhance intensities by up to three times; however, it also makes the relative intensities unreliable. For metal-carbonyl compounds the abundance problem can be alleviated by isotopic enrichment: in many cases metal-bound carbonyl groups will undergo exchange with gaseous ¹³CO.

(b) The nuclear relaxation time is often long, which means that long acquisition times are required. This can often be overcome by adding the paramagnetic complex $Cr(acac)_3$ to the solution (about 0.1 mol dm^{-3}), which increases the relaxation rate but appears to have little or no effect on chemical shifts. With this additive, however, the nuclear Overhauser effect is nullified; integrations again become useful, but there may be an increase in acquisition time.

(c) Many organometallic compounds are fluxional, so that fewer signals are seen than would be expected on the basis of the static structure. Cooling the solution may sometimes, but by no means always, slow the molecular contortions sufficiently for the limiting spectrum to be obtained. In extreme cases, it may be necessary to resort to the solid-state spectrum. Even then, π-bonded benzene or cyclopentadienyl rings may be rotating rapidly in their own planes, and still not show the expected inequivalences.

(d) Several metal nuclei have spin greater than 1/2. Carbon atoms directly bonded to n such atoms would be expected to show coupling, and to give $2nI+1$ lines. This is sometimes observed, but it is also possible that quadrupole effects will operate (see section 2.2.2) when the coupling is not resolved. A broad line then results, which may be difficult to observe. Coupling to $I=1/2$ nuclei (e.g. ¹⁰³Rh, ¹⁹⁵Pt) is usually seen and can be very informative.

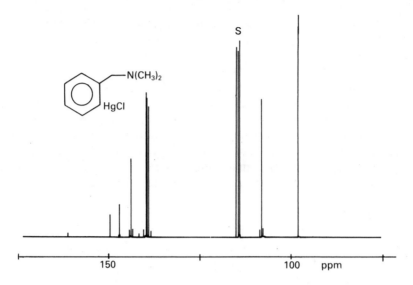

Fig. 2.20 — ^{13}C-{^{1}H} spectrum of HgCl(C$_6$H$_4$CH$_2$NMe$_2$). One signal is visible for each distinct type of carbon atom. The group marked S is due to the solvent, CDCl$_3$. The minor peaks are ^{199}Hg satellites (16.8%, $I=1/2$).

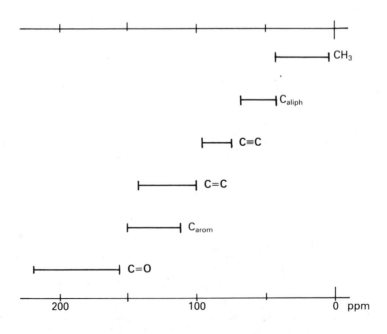

Fig. 2.21 — Characteristic chemical-shift ranges for ^{13}C in organic compounds.

Figure 2.20 shows the spectrum of a mercury(II) organometallic, in which separate signals for each type of carbon atom and satellites due to ^{199}Hg (16.8%, $I=1/2$) can be seen.

Except for the carbon atom directly bound to the metal atom, chemical shifts occur in the normal ranges (Fig. 2.21), which are well described in other, organic-chemistry based, texts. The signals for atoms immediately adjacent to the metal, however, usually appear to higher frequency than would otherwise be expected, sometimes considerably so. For instance, the C^1 resonance in σ-bonded phenyl groups occurs at 138.3 ppm in Ph_3P, and at 192.9 ppm in $Ph_2Ti(\pi\text{-}C_5H_5)_2$. Metal-bound methyl groups show a similarly large range: -38 to $+84$ ppm. A summary of the ranges for various types of organometallic ligand, including π-bonded groups, is given in Fig. 2.22.

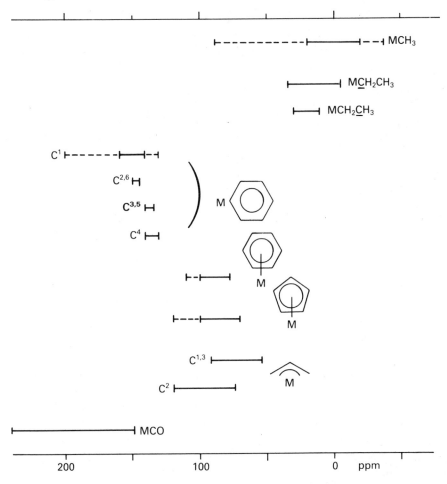

Fig. 2.22 — Characteristic ^{13}C chemical shift ranges for organometallic compounds. The majority of data for methyl and aromatic carbons lie in the major regions shown, but a few compounds give shifts in the dotted regions.

Greater systemization is possible in considering metal carbonyls. For a given type of compound, there is a systematic shift of the carbonyl resonance to lower frequency as the metal is changed from $3d$ to $4d$ to $5d$ (see Table 2.15). Bridging carbonyl groups occur at higher frequency than terminal groups. In substituted metal carbonyls, the group *trans* to the ligand usually has a higher chemical shift than that in the *cis* position. The chemical shift of the *trans* CO-group is also more sensitive to the identity of the substituent, and there appears to be a correlation with its σ/π-donor capacity. Both of these trends suggest that increased back-donation by the metal increases the carbonyl chemical shift. Similarly, as more carbonyl groups are replaced, so the chemical shift of those remaining increases.

Table 2.15 — ^{13}C chemical shifts for carbonyl groups (ppm)

	M=Cr		Mo		W	
M(CO)$_6$	211.2		200.7		191.4	
LM(CO)$_5$						
L=CH$_3$CN	219.2	213.9				
(PhO)$_3$P	217.5	213.8			197.0	194.5
(EtO)$_3$P			208.7	206.8	197.6	197.2
Ph$_3$P	221.3	216.5			199.8	198.0
Me(EtO)C	223.4	216.3			203.4	197.2
L$_2$M(CO)$_4$						
L=(MeO)$_3$P (*trans*)	219.3		210.3			
Et$_3$P (*trans*)					204.7	
(*cis*)					204.7	204.4
Ph$_2$P(CH$_2$)$_2$PPh$_2$			218.5	210.6		
norbornadiene	234.5	226.8	214.8	218.4	209.4	203.6
ArM(CO)$_3$						
Ar=mesitylene	235.1		223.7		212.6	
C$_6$(CH$_3$)$_6$	236.3		225.9		215.7	

Most data are for chloroform solutions.
Data from compilation by Todd and Wilkinson (see Bibliography).

As indicated earlier, ^{13}C measurements on carbonyl compounds are bedevilled by fluxionality effects. This may be due to straightforward librational motion, as in Fe(CO)$_5$ which shows only one ^{13}C resonance even at the lowest temperatures. (In fact, nearly all five-coordinate metal complexes are extremely flexible.) In poly-nuclear carbonyls, there is the additional complication of exchange between terminal and bridging positions. For instance, Fe$_3$(CO)$_{12}$ (structure **2.XV**) also shows only a single resonance at room temperature.

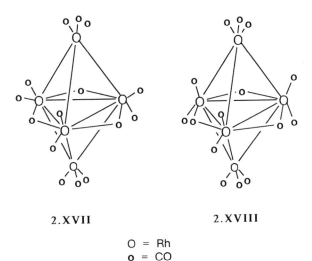

2.XV   2.XVI

A relatively simple example is $(\pi\text{-}C_5H_5)_2Rh_2(CO)_3$ which, in the solid state has one bridging and two terminal CO groups (2.XVI). At room temperature, the ^{13}C NMR spectrum shows a simple triplet (Fig. 2.23), showing that all CO groups are equivalent and couple equally to both metal atoms. On cooling the sample to $-80°$C, the spectrum is resolved into a doublet and a triplet, as would be expected from the solid-state structure.

Rhodium carbonyl compounds are widely used in homogeneous catalysis. For instance, the synthesis of ethylene glycol involves the anion cluster $Rh_5(CO)_{15}^-$. In the solid state, this cluster has the structure shown in **2.XVII**. Solutions can be studied only under high pressure of CO, and the ^{13}C NMR spectra obtained are shown in Fig. 2.24. The positions of the various carbonyl groups were determined by selective decoupling of the ^{103}Rh nuclei, whose chemical shifts are known from direct measurements. The spectra are simpler than required by the solid-state structure, which may indicate fluxionality (even at 201 K) or that the structure in solution is more symmetrical (e.g. as **2.XVIII**).

2.XVII   2.XVIII

O = Rh
o = CO

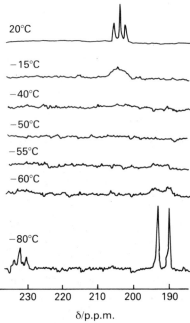

Fig. 2.23 — The ^{13}C-$\{^{1}H\}$ spectrum of the carbonyl region of $(\pi\text{-}C_5H_5)_2Rh_2(^{13}CO)_3$ at various temperatures. Reproduced with permission from J. E. Evans, B. F. G. Johnson, J. Lewis, T. W. Matheson, and J. R. Norton, *J. Chem. Soc., Dalton Trans,* (1978) 627.

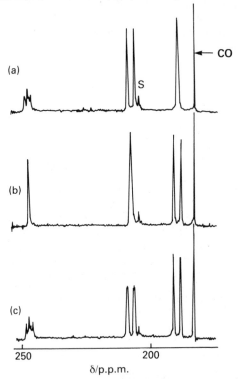

Fig. 2.24 — The ^{13}C-$\{^{1}H\}$ spectrum of $Rh_5(^{13}CO)_{15}^{-}$ under pressure of ^{13}CO (5 bar). In (a) and (b) the apical and equatorial ^{103}Rh nuclei respectively have been decoupled, (c) has no ^{103}Rh decoupling. Reproduced with permission from B. T. Heaton, L. Strona, J. Jonas, T. Eguchi, and G. A. Hoffman, *J. Chem. Soc., Dalton Trans.,* (1982) 1162.

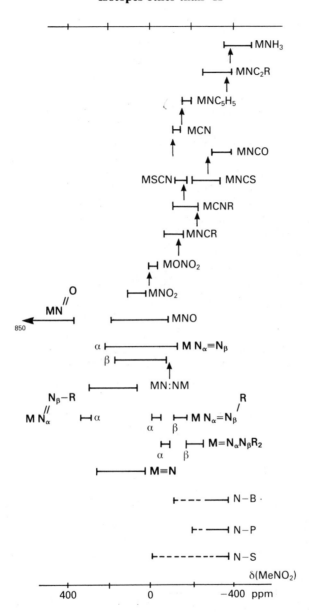

Fig. 2.25 — Chemical-shift scales for ¹⁴N and ¹⁵N in inorganic systems. For the metal complexes, arrows indicate the chemical shift of the free ligands (e.g. NH₃, MeNH₂, NC₅H₅, CN⁻, etc.). The dotted regions for boron, phosphorus, and sulphur derivatives refer to compounds with delocalized bonding. Adapted, with permission, from J. Mason, *Chem. Brit.*, (1983) 655.

2.4.3 Nitrogen-14 and Nitrogen-15

The nitrogen isotopes are finding increasing application, especially in biochemical studies. There are two isotopes which provide a rather invidious choice (Law of Inanimate Malice (Murphy)). For both isotopes, intrinsic sensitivity is low. ^{14}N has high natural abundance, 99.64%, but is quadrupolar ($I=3/2$) and often suffers from rapid relaxation, giving broad lines (see section 2.2). On the other hand, ^{15}N has spin of 1/2 but only 0.36% natural abundance, and its nuclear Overhauser factor is negative (section 2.4), so that proton decoupling gives reduced or even negative intensities. It also suffers from slow relaxation, necessitating long accumulation times. However, ^{15}N is becoming the isotope of choice, normally being used in wide-bore tubes (*ca.* 20 cm^3 volume), often with a relaxation agent such as Cr(acac)$_3$ to speed up acquisition. The use of a high-frequency spectrometer improves the intensities. When N–H coupling is expected, the INEPT technique is useful (see section 2.5.3). Good reviews of the subject covering both isotopes are available and should be consulted (see Bibliography).

Both isotopes give the same scale of chemical shifts, but this is another case in which there is confusion of sign conventions in the literature. There has also been a wide variety of materials used as standards. Aqueous ammonium or nitrate ions have been widely used, although their chemical shifts are somewhat dependent on concentration and pH. Neat nitromethane seems now to be gaining acceptance, with respect to which NH$_4^+$ has a shift of about -360 ppm.

The range of observed chemical shifts for the nitrogen isotopes is about 1000 ppm (Fig. 2.25). Co-ordination chemical shifts are quite small, 20–20 ppm, and may be

Table 2.16 — Data for ^{15}NH$_3$ in platinum(II) ammine complexes, showing the *trans* influence

	δ(MeNO$_2$)/ppm			^{1}J(N–Pt)/Hz		
trans to	N	Cl	O	N	Cl	O
cis-[PtCl$_2$(NH$_3$)$_2$]		422.1			325	
cis-[PtCl(H$_2$O)(NH$_3$)$_2$]$^+$		420.2	443.2		343	368
cis-[Pt(H$_2$O)$_2$(NH$_3$)$_2$]$^{2+}$			440.2		388	
[PtCl$_3$(NH$_3$)]$^-$		420.4			324	
[Pt(H$_2$O)$_3$(NH$_3$)]$^{2+}$			442.4			387
[PtCl(NH$_3$)$_3$]$^+$	420.4	424.2		282	329	
[Pt(NH$_3$)$_4$]$^{2+}$	421.2			286		
cis-[PtCl(NH$_3$)$_2$(py)]$^+$	423.5	417.6		273	343	
cis-[Pt(H$_2$O)(NH$_3$)$_2$(py)]$^{2+}$	421.3		436.8	290		384
cis-[Pt(NH$_3$)$_2$(py)$_2$]$^{2+}$	419.6			288		
[Pt(NH$_3$)(py)$_3$]$^{2+}$	414.7			302		
[Pt(NH$_3$)$_3$(py)]$^{2+}$	416.1			295		
cis-[PtCl(an)(NH$_3$)$_2$]$^+$	423.1	418.4		287	351	
cis-[Pt(an)$_2$(NH$_3$)$_2$]$^{2+}$	418.4			307		

an=C$_6$H$_5$NH$_2$,   py=C$_5$H$_5$N

Data from M. Nee and J. D. Roberts, *Biochem.*, **21** (1982) 4920.

positive or negative. Simple ligands such as NH_3, NO_2^-, etc., can be recognized readily, and there are characteristic differences between N- and S-bonded thiocyanate and between linear and bent metal-nitrosyl systems which are very useful. The ability to examine co-ordinated dinitrogen and to follow the effects of protonation and reduction makes the technique valuable in the study of potential nitrogen-fixation systems.

The simplest ligand, NH_3, occurs in a wide variety of complexes, and it is becoming possible to see the operation of the *trans*-influence; most data are available for platinum complexes, which show that as the *trans* ligand exerts a greater influence so $\delta(HN_3)$ and $^1J(N–Pt)$ decrease (Table 2.16).

There are also many nitrogen-containing inorganic compounds which do not involve metal atoms, such as the borazines, phosphazines and sulphur–nitrogen derivatives. The nitrogen chemical shifts observed in these materials depend on the bonding, being negative in saturated compounds like $R_3N.BR_3$ and becoming more positive for multiply-bonded and aromatic systems.

2.4.4 Fluorine-19

Fluorine is the least-used of the halogens, at least in inorganic chemistry. However, it is also the only halogen with which NMR is both simple and routinely practised. Fluorine-19 is 100% abundant, has $I=1/2$ and a high receptivity. Its NMR spectroscopy is thus very similar to that of ^{1}H, except that a slightly lower frequency is used (e.g. about 94 MHz), and that there is a very much wider range of chemical shifts (about 900 ppm). The latter phenomenon means that second-order effects are relatively rare. Fluorine can also appear in a bridging as well as a terminal position.

As with several other isotopes, care needs to be taken when reading the literature to note the chemical-shift standard employed. Most of the recent data are referred to $CFCl_3$, but CF_3COOH (-78.5 ppm relative to $CFCl_3$), C_6F_6 (-162.9 ppm), HF ($+198.4$ ppm) and even F_2 ($+422.9$ ppm) have been used. A change in sign-convention occurred around 1970–75, and data from this period or earlier may have the opposite signs from that employed today.

The chemical-shift ranges for organic compounds are summarized in Fig. 2.26. As a rough rule, increase in the electronegativity of the group to which fluorine is bonded results in a more positive (down-field) chemical shift. This trend is also noticeable in data for CF_3-groups summarized in Fig. 2.27.

Inorganic derivatives show rather wider variation in chemical shift, owing largely to the variety of electronegativities possible for the partner atoms, but shielding effects can be important for the heavier atoms. Data for the fluorides of most elements, and their derivatives, fall in the range -200 ppm (relative to $CFCl_3$) for p-block elements in moderate oxidation states to *ca.* 230 ppm for transition metals in high oxidation states (Fig. 2.28). Some typical data are given in Tables 2.17–2.19 Note that several factors other than electronegativity may affect chemical shifts.

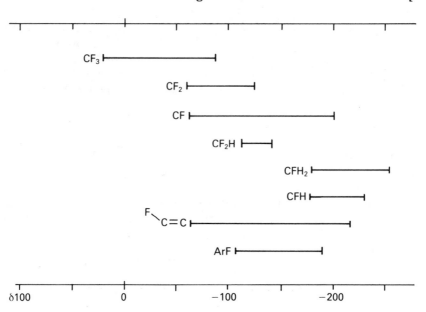

Fig. 2.26 — ^{19}F chemical-shift ranges for organofluorine derivatives.

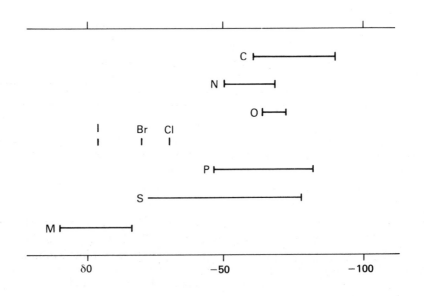

Fig. 2.27 — Chemical-shift ranges for ^{19}F in CF$_3$-groups bonded to various other atoms.

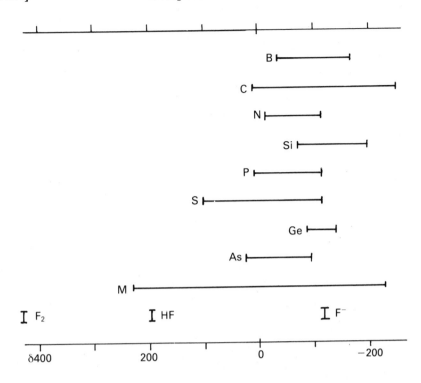

Fig. 2.28 — Chemical-shift ranges for fluorine directly bonded to other atoms.

Table 2.17 — ^{19}F data for organofluorophosphines

	Bu${}^t_2$PF	Bu${}^t_2$P(O)F	Bu^{t}P(O)F$_2$	Bu${}^t_3$PF$_2$	Bu${}^t_2$PF$_3$	Bu^{t}PF$_4$
δ(CFCl$_3$)	−202	−104	−84	−59	−48.6	−59.4
^{1}J(F–P)/Hz	848	1090	1195	808	908	1060

C. G. Dungan and J. R. van Wazer, *Compilation of Reported ^{19}F Chemical Shifts*, Wiley–Interscience, New York, 1970.

In many molecules, there are non-equivalent fluorine atoms present, which usually give distinct, well-resolved signals, and show F–F coupling of 30–80 Hz (see problems 2.15 and 2.16, and Table 2.18). To give a rather extreme example of the range of chemical shifts which may be encountered for a single compound, data for the fluorine-bridged compounds FXe(μ-F)MOF$_4$ (M=Mo, W) are illustrated in Table 2.20. Fluorine bound to xenon(II) has much lower chemical shifts than that bonded to molybdenum(VI) or tungsten(VI), and the bridging fluorine atom has an

Table 2.18 — ^{19}F data for tungsten(VI) chlorofluorides

Compound			δ(CFCl$_3$)	^{2}J(F–F)/Hz	^{1}J(F–^{183}W)/Hz
WF$_8^{2-}$			63.8		68
WF$_7^-$			134		
WF$_6$			164.5		44
WF$_5$Cl	axial		125.1	73	
	equatorial		180.5		25
WF$_4$Cl$_2$-*trans*			189.5		20
	-*cis*	axial	134.5	66	
		equatorial	197.5		
WF$_3$Cl$_3$-*fac*			142.5		
	-*mer*		137.3	62	
			207.5		
WF$_2$Cl$_4$-*trans*			218.5		
	-*cis*		150.5		
WFCl$_5$			156.5		

Table 2.19 — ^{19}F data for xenon fluorides (in HF). [Chemical-shift values vary by a few ppm between different reports]

Compound	δ(CFCl$_3$)	^{1}J(F–^{129}Xe)/Hz
XeF$_2$	−206	5600
XeF$_4$	−23	3860
XeF$_6$	+50	
XeOF$_4$	+101	1115
XeO$_2$F$_2$	+105	1124

Table 2.20 — Data for F^{a}Xe(μ–F^b)MOF$_4^c$

M	δa	δb	δc	J(Xe–a)	J(Xe–b)	J(a–b)	J(b–c)
Mo	−223.1	−170.0	141.8	6140	5117	264	50
W	−228.9	−168.8	135.8	6150	5016	275	50

intermediate shift. An increase in chemical shift for a bridging fluorine atom is normal: the cation in $[Xe_2F_3][AsF_6]$ shows signals at 252 ppm (terminal) and -185 ppm (bridging) $[^2J(F-F)$ is 308 Hz].

2.4.5 Aluminium-27
NMR measurements with ^{27}Al are useful in a variety of systems (see Fig. 2.29), but

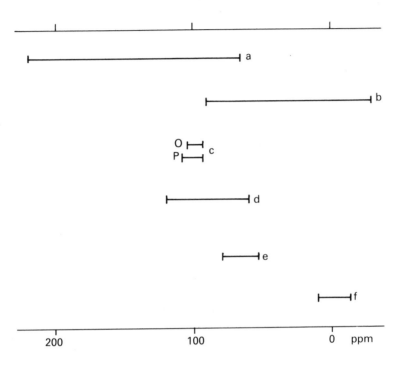

Fig. 2.29 — ^{27}Al chemical shifts, relative to $Al(H_2O)_6^{3+}$. The ranges marked correspond to: a, alkyls, hydrides, and their adducts; b, halides and AlX_4^-; c, halide adducts with O- or P-donor ligands; d, alkoxides and their adducts; e, tetrahedral AlO_4 co-ordination; f, octahedral AlO_6 co-ordination in hydrates, hydroxides, aluminates, zeolites.

interpretation is frequently complicated by large line widths caused by quadrupolar relaxation. ^{27}Al has $I=5/2$ and a relatively large quadrupole moment, and relaxation is very efficient in any system with less than cubic symmetry (section 2.2.2). In strict octahedral or tetrahedral geometry the line width may be only a few hertz but is more usually several hundred hertz (the record is currently 60 kHz). This means that coupling to other nuclei is not usually resolved: major exceptions are some hydride systems, phosphine complexes of the halides, and some solvated species such as $[Al(dmmp)_n(H_2O)_{6-n}]^{3+}$ $[n=1,2;$ dmmp$=$MePO(OMe)$_2]$. Similarly, coupling to ^{27}Al is rarely observed in the spectra of other nuclei. There is much current interest in the use of the CPMAS technique to investigate the spectra of solids, especially synthetic aluminosilicate catalysts. It is easy to distinguish tetrahedrally co-ordinated aluminium in the framework (δ 60–80) from octahedrally co-ordinated interstitial or surface species (δ 0–20). In conjunction with the ^{29}Si spectrum, a good picture of the structure can be obtained, and its changes during reactions followed.

2.4.6 Silicon-29
In marked contrast to aluminium, ^{29}Si has $I=1/2$ and gives quite sharp NMR spectra. However, the basic sensitivity is low and this, combined with an abundance of only 4.7%, makes the receptivity poor. Relaxation is also slow, and spectrum accumulation can be tedious. The magnetogyric ratio is negative, so that ^{1}H decoupling can lead to weak, zero, or negative signals through NOE (section 2.4). Considerable improvement in accumulation time and line intensity can be achieved by the use of spin polarization techniques (e.g. INEPT or DEPT — section 2.5.3). Nevertheless, this isotope is becoming increasingly popular, especially for studies on polymeric siloxanes and silicates, for which structural information is almost impossible to obtain by any other method.

Typical ranges of chemical shifts are shown in Fig. 2.30. In general, signals are

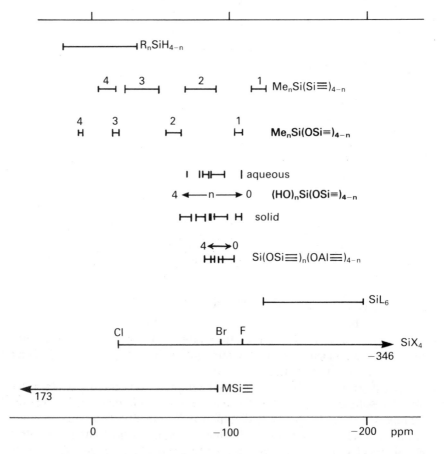

Fig. 2.30 — ^{29}Si chemical shift ranges, relative to Si(CH$_3$)$_4$. The lowest line refers to transition-metal silyl compounds.

well separated and even closely related species can be differentiated. For example, in polysiloxanes, distinction between silicon atoms carrying different numbers of organic groups, $R_nSi(OSi\equiv)_{4-n}$, is simple, as would be expected. As the number of electronegative oxygen neighbours increases, the resonance moves to lower frequency. Small differences are also seen between silicon atoms with the same number of organic groups but different neighbouring groups (see Fig. 2.31). Similar differen-

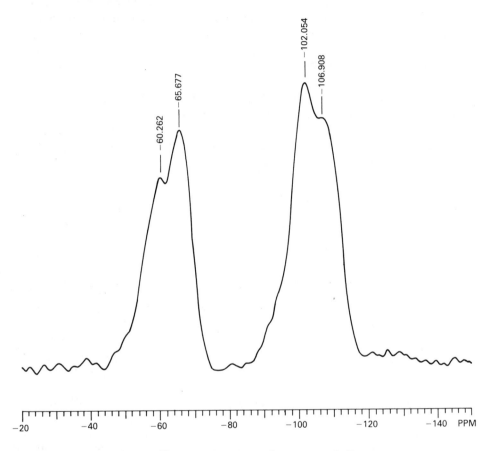

Fig. 2.31 — ^{29}Si CPMAS spectrum of an organo-polysiloxane.

tiation is found in silicate anions in solution, and in silica gels and solid silicates; in solid aluminosilicates, it is even possible to distinguish SiO_4 groups with different numbers of aluminium neighbours.

An increasing number of transition-metal complexes is becoming available which contain metal–silicon bonds. While these are normally characterized by ^{1}H or ^{31}P measurements, ^{29}Si is also being practised. The chemical shift of the co-ordinated silyl group is usually to low field of the corresponding methyl silane, e.g. for $[Fe(SiMe_3)(\pi\text{-}C_5H_5)(CO)L]$ the recorded shifts are 41.3 ppm for L=CO and 34.75 for L=PPh$_3$, whereas MeSiMe$_3$ is the reference material, with zero shift.

2.4.7 Transition metals

All the transition metals have at least one isotope which could, in principle, be examined by NMR. Many have low receptivities or large quadrupole moments, which make their observation difficult (see Appendix 1). Significant work has been done with ^{51}V, ^{93}Nb, ^{183}W, ^{55}Mn, ^{59}Co, ^{103}Rh and ^{195}Pt.

The chemical shifts cover very wide ranges, but all follow broadly similar trends: decrease in oxidation state and increase in softness of the ligands give increased shielding and shifts to low frequency. Data for the Group VIII metals are summarized in Fig. 2.32 and some representative data are given in Table 2.21. In addition to the gross changes produced by varying the ligands, there are also discernible differences in chemical shift between various isomers.

For ^{59}Co, quadrupole relaxation produces line broadening when the symmetry is low, and there is a good correlation with the expected magnitude of the electric-field gradient (EFG) calculated by the point-charge model (see section 4.3.3). The EFG is zero for octahedral MA_6 and fac-MA_3B_3 geometry. For other mixed-ligand complexes it is expected to increase as the difference between the ligands increases, and to vary with geometry in the order $MA_5B{\sim}cis$-$MA_4B_2{<}mer$-$MA_3B_3{<}trans$-MA_4B_2. Note that the lines can be very broad indeed, e.g. cis- and $trans$-$[Coen_2Cl_2]^+$ have widths of 2.0 and 4.0 G, which are equivalent to 340 and 600 ppm; it is then impossible to determine the chemical shift with any great accuracy.

Most of the other isotopes are quite well behaved, the principal problems being low receptivity or difficult frequency ranges.

2.5 SPECIAL METHODS — MULTIPLE RESONANCE TECHNIQUES

A variety of rather complicated special techniques is available to assist in the unravelling of complex spectra. They are all rather exotic and lie beyond the scope of the present treatment, since they involve the application of two or more frequencies simultaneously and special pulse sequences; their use requires expert assistance. A brief description of some of the more important methods is given below, and some are illustrated by application to $(CH_3CH_2O)_2P(O)CH_2CH_2Br$. The conventional ^1H and ^{13}C spectra of this compound are shown in Fig. 2.33 (p. 88). Although it is evident from normal chemical-shift and integration considerations that ^1H signals a and d correspond to the CH_2 and CH_3 of the ethoxy group, the assignment of the remaining CH_2 groups is far less obvious. In the ^{13}C spectrum, all signals are doubled by coupling to ^{31}P, although only that in q is well resolved. This probably (but by no means certainly) means that C_q is bonded directly to the phosphorus atom.

2.5.1 Two-dimensional spectra

FT spectra are accumulated by applying a series of broad pulses and performing a Fourier transform on the free-induction decay (FID) signals. A variable delay is introduced before the FID is sampled by adding a second series of trigger pulses, which effectively applies a phase difference to the spectrum, i.e. it offsets the frequency slightly, which alters the shape of the spectrum. This provides the second dimension to the spectrum. A series of spectra is obtained at different trigger-pulse frequencies. The nett result is as shown for the model compound in Fig. 2.34, where

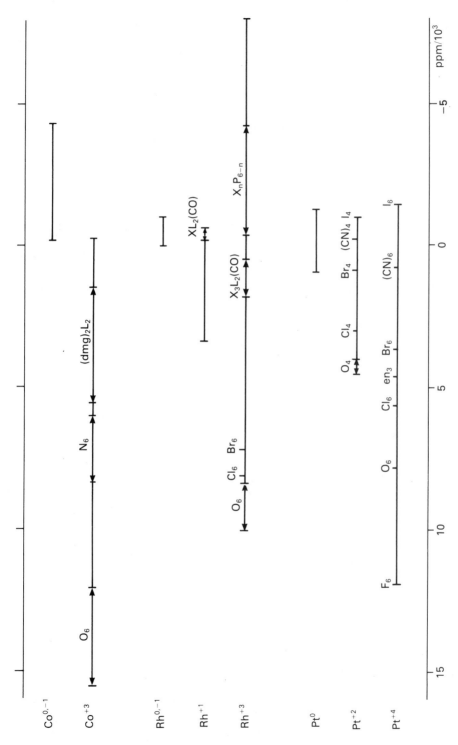

Fig. 2.32 — Chemical-shift ranges for ^{59}Co, ^{103}Rh and ^{195}Pt in various oxidation states. The ranges for oxidation state zero include cluster compounds. Note that in every case, the softest ligands give the most negative chemical shifts.

Table 2.21 — Data for cobalt and rhodium complexes

$^{59}Co^{III}$	$\delta^{a,b,c}$	$^{103}Rh^{III}$	$\delta^{d,e}$
$[Co(NH_3)_6]^{3+}$	8107(0.05)	$[RhCl_4(AsMe_3)_2]^-$-*trans*	4735
$[CoCl(NH_3)_5]^{2+}$	8807(1.0)	$[RhBr_4(AsMe_3)_2]^-$-*trans*	3950
$[CoCl_2(NH_3)_4]^+$-*cis*	9770(2.0)	$[RhI_4(AsMe_3)_2]^-$-*trans*	2044
-*trans*	9770(4.0)	$[RhCl_3(AsMe_3)_3]$-*fac*	2125
$[Co(NH_3)_5N_3]^{2+}$	8637(0.2)	$[RhBr_3(AsMe_3)_3]$-*fac*	1798
$[Co(NH_3)_4(N_3)_2]^+$-*cis*	9217(0.21)	$[RhCl_3(AsMe_3)_3]$-*mer*	2806
-*trans*	9177(0.30)	$[RhBr_3(AsMe_3)_3]$-*mer*	2276
$[Co(NH_3)_3(N_3)_3]$-*mer*	9857(0.71)	$[RhI_3(AsMe_3)_3]$-*mer*	1191
-*fac*	9927(0.17)	$[RhCl_2(AsMe_3)_4]^+$-*cis*	1457
$[Coen_3]^{3+}$	7077(0.05)	$[RhBr_2(AsMe_3)_4]^+$-*cis*	1222
$[Coen_2(C_2O_4)]^+$	8567(2.0)	$[RhCl_2(AsMe_3)_4]^+$-*trans*	1946
$[Coen(C_2O_4)_2]^-$	10357(3.2)	$[RhBr_2(AsMe_3)_4]^+$-*trans*	1488
$[Co(C_2O_4)_3]^{3-}$	12807(0.25)	$[RhI_2(AsMe_3)_4]^+$-*trans*	658

[a] Figures in parentheses are line widths in gauss (G=*ca* 1020 Hz=*ca* 75 ppm).
[b] Relative to $[Co(CN)_6]^{3-}$.
[c] Data from F. Yajima, Y. Koike, A. Yamasaki and F. Fujiwara, *Bull. Chem. Soc. Japan*, **47**, (1974) 1442.
[d] Relative to $[Rh(CN)_6]^{3-}$.
[e] Data from R. J. Goodfellow, in J. Mason, ed. *Multinuclear NMR* (Plenum Press, New York, 1987), Chapter 20.

Table 2.22 — Data for platinum complexes

	δ^a	^1J(Pt–N)/Hz		
	trans to	Cl	NH$_3$	dmso
[PtCl$_2$(NH$_3$)$_2$]-*cis*	2097	312		
[PtCl(NH$_3$)$_3$]$^+$	2354	317	278	
[PtCl$_2$(dmso)(NH$_3$)]-*cis*	3046	336		
[PtCl$_2$(dmso)(NH$_3$)]-*trans*	3067			273
[PtCl(dmso)(NH$_3$)$_2$]$^+$-*trans*	3126		287	
[PtCl(dmso)(NH$_3$)$_2$]$^+$-*cis*	3147	340		234
[Pt(dmso)(NH$_3$)$_3$]$^{2+}$	3224		288	232

dmso$=$(CH$_3$)$_2$SO

^{a}Relative to Na$_2$[PtCl$_6$]/D$_2$O.
Data from S. J. S. Kerrison, and P. J. Sadler, *J. Chem. Soc., Chem. Comm.* (1977) 861.

each nucleus is seen at its correct chemical shift but the ^{1}H–^{1}H coupling pattern is displayed vertically. This is particularly valuable when the normal spectrum is heavily overlapped. (It is usual also to show the intensities of the various signals, either by contour mapping or by making a three-dimensional representation.) Note that, for (EtO)$_2$P(O)CH$_2$CH$_2$Br, each ^{1}H signal is doubled by coupling to the ^{31}P nucleus; in this presentation, each half of each double set appears to have a different chemical shift. H$_c$ thus appears to show the greatest coupling to ^{31}P, which may indicate that this CH$_2$ group is the one bound directly to phosphorus; further confirmation would be desirable, however (see below).

A further development of this technique is **correlation spectroscopy** (COSY) in which both dimensions are chemical-shift scales. When both scales refer to the same isotope, a series of peaks is found along the main diagonal, i.e. the various nuclei A, B, C, etc., appear at the points (δ_A, δ_A), (δ_B, δ_B), (δ_C, δ_C), etc. However, if two nuclei are coupled, say A and B, extra signals are found at the off-diagonal positions (δ_A, δ_B) and (δ_B, δ_A) (see Fig. 2.35). This greatly simplifies the assignments. Thus, for (EtO)$_2$P(O)CH$_2$CH$_2$Br, confirmation is obtained that H$_a$ and H$_d$ are coupled together (ethoxy groups) as are H$_b$ and H$_c$ (bromoethyl group). (Note also that the lack of coupling confirms that signal h belongs to an impurity.)

It is also possible to obtain a two-dimensional plot of chemical shifts for two different isotopes, e.g. ^{1}H versus ^{13}C, which allows the two sets of signals to be correlated, again aiding the assignment. For the model compound (Fig. 2.36), it is now evident that H$_d$ is associated with C$_s$, confirming that these constitute the methyl group. Similarly, H$_c$ is linked to C$_q$; that is, both atoms with the largest coupling to ^{31}P are part of the same CH$_2$ group, presumably that bonded to the phosphorus.

2.5.2   INDOR — Internuclear Double Resonance
For a spin-coupled system involving different isotopes, it is possible to obtain the line positions for one isotope (say, X) while observing the other (say, A). This can be done (and is most useful) even when the X-spectrum cannot be obtained directly, for

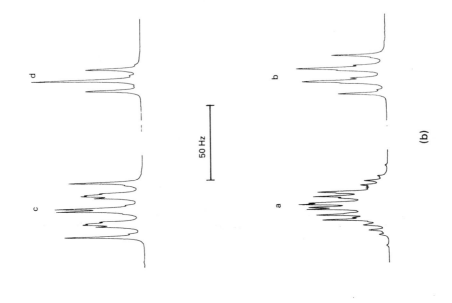

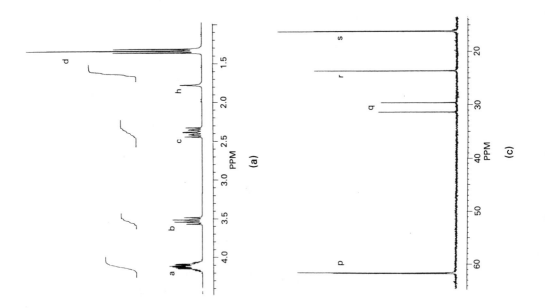

Fig. 2.33 — NMR spectra of $(C_2H_5O)_2P(O)CH_2CH_2Br$. (a) ^1H spectrum at 300 MHz. (b) ^1H resonances on an expanded scale. (c) ^{13}C-$\{^1H\}$ spectrum at 75 MHz.

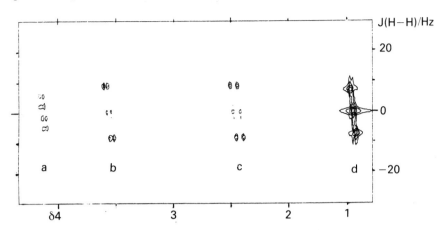

Fig. 2.34 — Two-dimensional, J-resolved ^{1}H spectrum of $(C_2H_5O)_2P(O)CH_2CH_2Br$. Note that only the ^{1}H–^{1}H coupling is resolved vertically; the ^{1}H–^{31}P coupling appears as a splitting of the chemical shift.

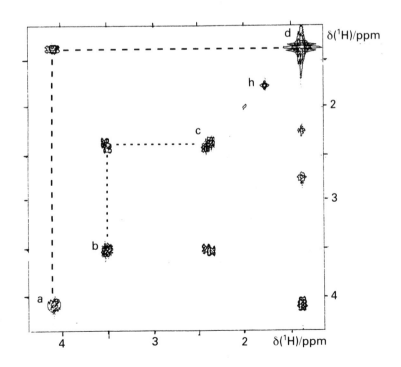

Fig. 2.35 — COSY (^{1}H–^{1}H) spectrum for $(C_2H_5O)_2P(O)CH_2CH_2Br$. The off-diagonal peaks indicate that the protons are coupled in the pairs a, d and b, c.

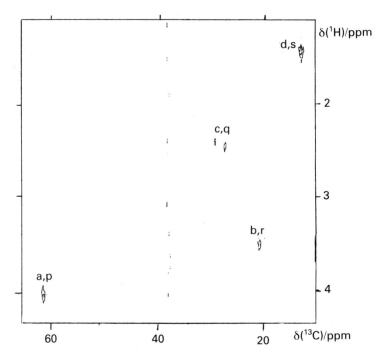

Fig. 2.36 — COSY (^{13}C–^1H) for $(C_2H_5O)_2P(O)CH_2CH_2Br$. The positions of the peaks indicate which atoms are associated, i.e. proton a with carbon p, etc.

example when the lines are very broad. The technique involves applying two frequencies simultaneously, one to excite a particular resonance of the A-spectrum and the other to sweep the X-spectrum. Each time an X-resonance frequency is reached for an X-nucleus which is coupled to the A-nucleus under observation, the intensity of the A-signal changes. Thus, a representation of the X-spectrum can be built up. In this way it has been possible, for instance, to obtain spectra for isotopes with inconvenient frequency ranges, e.g. ^{14}N or ^{103}Rh, by using the ^1H resonance.

In the COSY technique, the X-frequencies are applied at quite low power. If high power is used, the X-resonances become saturated and the A–X coupling is lost; this is the principle of spin decoupling. An intermediate power level is sometimes used, when it becomes possible to determine the signs of the coupling constants in three-spin AMX systems; the method is then known as 'spin tickling'.

An extension of this method is used in solid-state MAS spectroscopy, since the application of the second frequency can result in enhancement of the intensity of the observed spectrum by allowing energy to be transferred between the nuclei of different isotopes, e.g. from ^{13}C to ^1H. Such a process is known as 'cross polarization', resulting in the acronym CPMAS.

2.5.3 DEPT — Distortionless Enhancement by Polarization Transfer, and INEPT — Insensitive Nuclei Enhancement by Polarization Transfer

These methods utilize the nuclear Overhauser effect to enhance intensities. Carefully timed secondary pulses covering the full range of X-frequencies are applied before acquisition of the A-spectrum, which can result in increases in intensity by factors up to γ_X/γ_A. At the same time the shape of the spectrum changes drastically, and careful analysis is required. It is also possible by this technique to simplify complicated spectra by giving a negative intensity to signals with particular coupling patterns (Fig. 2.37). With special programming, it is possible to suppress all signals

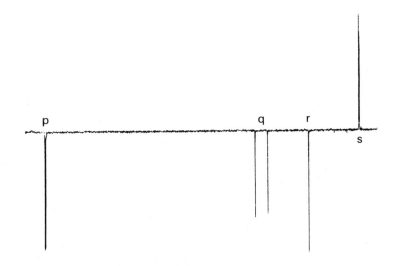

Fig. 2.37 — ^{13}C-DEPT135° spectrum for $(C_2H_5O)_2P(O)CH_2CH_2Br$. In this presentation, CH_2-groups appear with negative intensity.

other than those with particular couplings. For instance, in a ^{13}C spectrum, the signals for all CH_2 groups could be selected and all other resonances ignored. Equally, it is possible to observe just a satellite subspectrum and to suppress the major signal.

2.5.4   DANTE — Delays Alternating with Nutations for Tailored Excitation
If pulses at a single resonance frequency are applied with the decoupler in operation, the spectrum will consist of a single line corresponding to the chemical shift of a particular nucleus. By carefully controlling the timing of the operation of the decoupler, it is possible to observe the full coupling pattern associated with this nucleus. A complex spectrum can thus be simplified into its component subspectra.

BIBLIOGRAPHY

J. W. Akitt, *NMR and Chemistry. An Introduction to the Fourier Transform — Multinuclear Era* 2nd edn., Chapman & Hall, London, 1983. Somewhat theoretical.

E. A. V. Ebsworth, D. W. H. Rankin, and S. J. Cradock, *Structural Methods in*

Inorganic Chemistry. Blackwell Scientific, Oxford, 1987. Chapter 2 deals with NMR and gives some quite complex examples.

J. Mason (Ed.), *Multinuclear NMR.* Plenum Press, New York, 1987. General chapters on parameters and relaxation. Specific chapters on the various Groups of the Periodic Table.

R. K. Harris and B. E. Mann, *NMR and the Periodic Table.* Academic Press, London, 1978. A useful introductory survey; somewhat dated.

P. Laszlo (Ed.), *NMR of Newly Accessible Nuclei.* Academic Press, New York, 1983. Two volumes. Volume 1 deals with instrumentation and theoretical aspects at a fairly advanced level and some specific systems including halogens other than fluorine. Volume 2 has specific chapters on ^{2}D; ^{3}T; ^{11}B; ^{17}O; Alkali Metals; ^{27}Al; ^{29}Si; ^{25}Mg, ^{43}Ca; ^{59}Co; ^{77}Se, ^{125}Te; ^{103}Rh; ^{109}Ag; ^{113}Cd; ^{203,205}Tl; less common nuclei.

B. E. Mann, *Adv. Organometal. Chem.,* **28** (1988) 398. 'Recent Developments in NMR Spectroscopy of Organometallic Compounds'. Survey of modern experimental methods.

R. H. Crabtree and D. G. Hamilton, *Adv. Organometal. Chem.,* **28** (1989) 299. 'H–H, C–H and Related Sigma-bonded Groups as Ligands'. Dihydrogen and 'agostic' C–H ligands.

L. J. Todd and J. R. Wilkinson, *J. Organomet. Chem.,* **77** (1974) 1. 'Carbon-13 NMR Spectra of Metal–Carbonyl Compounds'. Contains much data.

B. E. Mann, *Adv. Organometal. Chem.,* **12** (1974) 135. 'Carbon-13 Chemical Shifts and Coupling Constants of Organometallic Compounds'. Much data tabulated.

J. Mason, *Chem Rev.,* **81** (1981) 205. 'Nitrogen Nuclear Magnetic Resonance Spectroscopy in Inorganic, Organometallic, and Bioinorganic Chemistry'.

J. Mason, *Chemistry in Britain,* **19** (1983) 654. 'Patterns and Prospects in Nitrogen NMR'. Fairly general.

M. Witanowski, L. Stefaniak, and G. A. Webb, *Ann. Reports NMR Spectroscopy,* **11B**, (1981) 2. 'Nitrogen NMR Spectroscopy'. The technique and data.

P. S. Pregosin and R. W. Kunz, *^{31}P and ^{13}C NMR of Transition Metal Phosphine Complexes.* Springer-Verlag, Berlin, 1979. A good survey.

G. N. La Mar, W. de W. Horrocks, and R. H. Holm, *NMR of Paramagnetic Molecules: Principles and Applications.* Academic Press, 1973.

H. Sigel (ed.), *Biological Magnetic Resonance,* **21** (1987). 'Applications of NMR to Paramagnetic Species'.

Annual Reports of NMR Spectroscopy contain many useful articles, of which the following may be of interest.

P. S. Pregosin, **11A** (1981) 227. '^{13}C NMR of Group VIII Metal Complexes'. Much data for organometallic, carbonyl and other coordination compounds.

E. A. Williams, **15** (1983) 235. 'Recent Advances in Silicon-29 NMR Spectroscopy'.

T. Drakenberg, **17**, (1986) 231. 'Nuclear Magnetic Resonance of less Common Quadrupolar Nuclei'. Includes gallium, arsenic, antimony and transition metals.

P. S. Pregosin, **17** (1986) 285. 'Platinum NMR Spectroscopy'.

A. R. Seidle, **12** (1982) 177. 'Boron-11 NMR Spectroscopy'.

R. E. Waylishen and C. A. Fyfe, **12** (1982) 1. 'High-Resolution NMR of Solids'.

W. McFarlane and D. S. Rycroft, **16** (1985) 293. 'Multiple Resonance'.

B. E. Mann, **12** (1982) 263. 'Dynamic NMR Spectroscopy in Inorganic and Organometallic Chemistry'.

K. G. Orrell and V. Sik, **19** (1987) 79. 'Dynamic NMR Spectroscopy in Inorganic and Organometallic Chemistry'.

Data compilations

Searchable compilations of ^{13}C- and ^{19}F-data are available (mainly for organic compounds, but including organometallics):

STN C13NMR: through Scientific and Technical Information Network, PO Box 2465, D-7500 Karlsruhe, 1, Germany (UK agents, the Royal Society of Chemistry) 19F: Preston Scientific Ltd., Blackburn, Lancashire, UK.

PROBLEMS

(Answers will be found on pp. 104–107)

P2.1   Fig. 2.38 gives a representation of a product described as C_3H_5PdI. Does this contain σ- or π-bonded allyl groups?

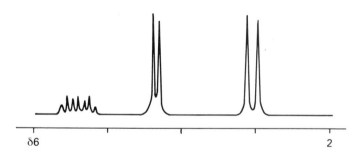

δ6   2

Fig. 2.38.

P2.2   Sketch the 100 MHz spectra expected for the high-field region of the following complexes (some of which are hypothetical). Treat all possible geometrical isomers, assume that arsenic, iridium, and palladium do not couple to other nuclei.

Take $^1J(Pt–H)$ as 800 Hz; $^1J(Rh–H)$ as 12 Hz; $^2J(P_{trans}–H)$ as 120 Hz; $^2J(P_{cis}–H)$ as 20 Hz (if two distinct *cis* phosphines are present, take 16 and 20 Hz). H–H coupling, $^2J(H–H)$ 2–5 Hz, would not be resolved.

(a) $IrHCl_2(AsMePh_2)_3$ (b) $IrH_2Cl(AsMePh_2)_3$
(c) $RhHCl_2(AsMePh_2)_3$ (d) $RhH_2Cl(AsMePh_2)_3$
(e) $IrHCl_2(CO)(PMePh_2)_2$ (f) $IrH_2Cl(CO)(PMePh_2)_2$
(g) $PdH(CN)(PPh_3)_2$ (h) $PtH(CN)(PPh_3)_2$

P2.3   The 100 MHz ^1H spectra given in Fig. 2.39 were unfortunately not labelled when they were run, but are known to correspond to the complexes listed. Assign each spectrum to the appropriate complex, giving your reasons.

(a) *trans*-PtHBr(PEt_3)_2   (b) PdH(CN)(Ph_2PCH_2CH_2PPh_2)
(c) [PtH(PEt_3)_3]ClO_4   (d) *trans*-PdHCl(PEt_3)_2

For the two platinum complexes, estimate the value of $^1J(H–Pt)$.

P2.4   Does the 200 MHz ^1H spectrum shown in Fig. 2.40 correspond to the *fac* or the *mer* isomer of $IrHCl_2(PMe_2Ph)_3$? Which ligand is *trans* to the hydride?

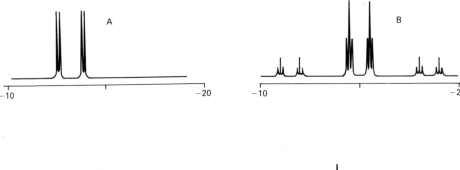

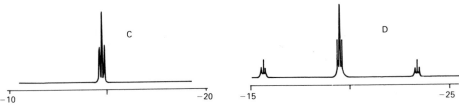

Fig. 2.39

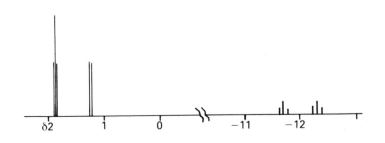

Fig. 2.40.

P2.5 Use the information given to identify the complexes *E* and *F*, to make estimates of their NMR parameters, and to deduce their molecular structures.

The complex [PtCl$_2$(PEt$_3$)]$_2$ reacts with triethylphosphine in non-polar solvents to form the pale yellow complex *E*. Complex *E* reacts in ethanol with hydrazine to give a white complex, *F*. Representations of the ^{31}P–{^{1}H} NMR spectra of *E* and *F* are shown in Fig. 2.41 together with part of the ^{1}H spectrum of *F*. The numbers on the last spectrum are the frequencies of the individual lines.

P2.6 From the data given below, suggest structures for the complexes MCl$_3$L$_3$ and the two forms of MHCl$_2$L$_3$ (M=Rh, Ir; L=tertiary phosphine or arsine).

When hydrated iridium trichloride is treated in the cold with a tertiary phosphine or arsine (L), white products of formula [IrCl$_3$L$_3$] are obtained. When these are

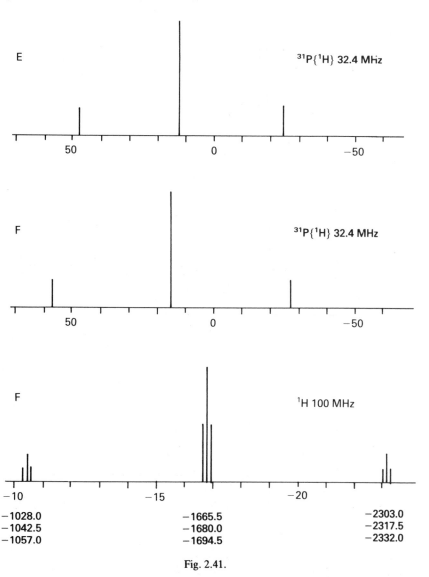

Fig. 2.41.

refluxed with ethanolic potassium hydroxide, yellow monohydride complexes are formed, [IrHCl$_2$L$_3$]. However, different isomers are obtained, depending on the duration of the reaction: form (I) is produced after about one hour, whereas prolonged reaction gives form (II). When solutions of form (I) are heated or exposed to light, isomerization to form (II) occurs. Rhodium forms a completely analogous series of complexes. ^1H spectroscopic data are given below, from which the structures may be deduced, and some of the spectra are illustrated (Fig. 2.42, 100 MHz). [You may assume that all complexes of a given type give analogous spectra, even though full sets of data are not available.] Coupling is *not* observed with arsenic or with iridium, but may be seen for M=Rh (100%, I=1/2).

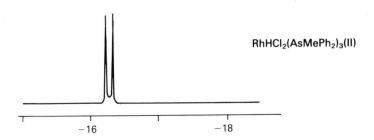

RhHCl$_2$(AsMePh$_2$)$_3$(II)

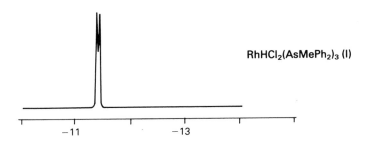

RhHCl$_2$(AsMePh$_2$)$_3$ (I)

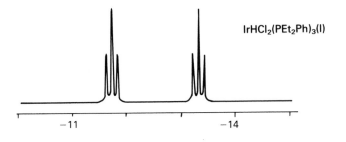

IrHCl$_2$(PEt$_2$Ph)$_3$(I)

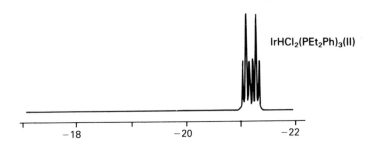

IrHCl$_2$(PEt$_2$Ph)$_3$(II)

Fig. 2.42 (cont. on pp. 97–98).

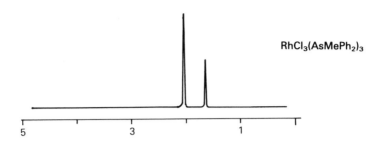

RhCl$_3$(AsMePh$_2$)$_3$

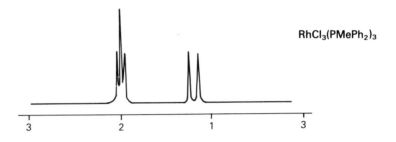

RhCl$_3$(PMePh$_2$)$_3$

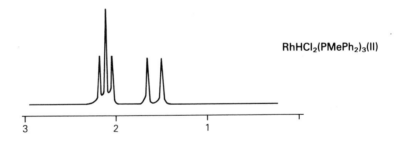

RhHCl$_2$(PMePh$_2$)$_3$(II)

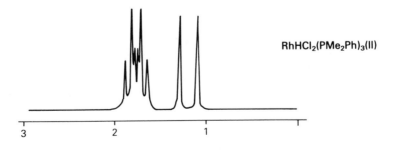

RhHCl$_2$(PMe$_2$Ph)$_3$(II)

Fig. 2.42.

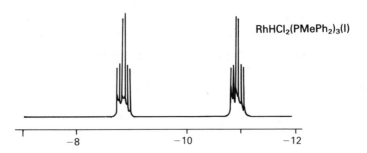

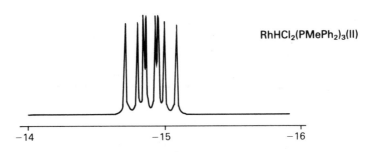

Fig. 2.42.

		$\delta(^1H)$	$^2J(H\text{–}P)/Hz$		$^1J(H\text{–}Rh)/Hz$
RhHCl$_2$(AsMePh$_2$)$_3$	(I)	−11.4			4
	(II)	−16.3			9
RhHCl$_2$(PMePh$_2$)$_3$	(I)	−9.9	206 ,	9	4
	(II)	−14.9	14.5,	9	13.5
IrHCl$_2$(PEt$_3$)$_3$	(I)	−12.6	163 ,	20	
	(II)	−21.6	15 ,	15	
IrHCl$_2$(PEt$_2$Ph)$_3$	(I)	−12.6	158 ,	19	
	(II)	−21.2	18 ,	12	

P2.7   For the complexes (e)–(h) of Problem 2.2, sketch the ^{31}P–{^{1}H} NMR spectra you would expect (at 30 MHz). Treat all possible isomers. Assume that the chemical shifts of non-equivalent phosphorus atoms are 10 ppm apart (in practice differences lie within ±10 ppm, and depend on the metal and the other ligands present).

Take $^1J(P\text{–}Pt)$ as 2000 Hz (*trans* to P), 2500 Hz (*trans* to H$^-$), or 2800Hz (*trans* to CN$^-$).

P2.8   Identify the product of the reactions described below.

(a) The aqueous solution obtained by reaction of *cis*-PtCl$_2$(PEt$_3$)$_2$ with one molar equivalent of Ag$_2$SO$_4$ followed by acidification with H$_2$SO$_4$ was extracted with

CHCl$_3$. The chloroform solution gave a ^{31}P–{^{1}H} NMR spectrum consisting of a single line at 4.93 ppm (H$_3$PO$_4$) with satellites separated by 3723 Hz.

(b) When the above reaction was repeated in the presence of Et$_3$P (one mol), the filtered aqueous solution showed two ^{31}P–{^{1}H} signals with satellites: 24.7 ppm (doublet, 19.5 Hz), 2265 Hz, and 3.77 ppm (triplet, 10.5 Hz), 3652 Hz.

P2.9 The ^{13}C–{^{1}H} spectrum shown in Fig. 2.43 is of the product obtained by

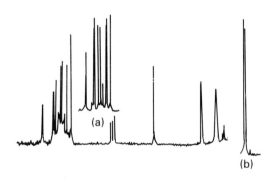

Fig. 2.43 — (Reproduced with permission from M. F. Lappert, P. I. Riley, P. I. W. Yarrow, J. L. Atwood, W. E. Hunter, and M. J. Zaworotoko, *J. Chem. Soc., Dalton Trans.*, (1981) 814).

treating [Zr(η^5-C$_5$H$_4$CH$_3$)$_2$Cl$_2$] with Li[CH(SiMe$_3$)$_2$]. Does the spectrum contain the resonances expected for [Zr(η^5-C$_5$H$_4$CH$_3$)$_2${CH(SiMe$_3$)$_2$}Cl]? [The set of three peaks in the middle portion of the spectrum is due to the solvent, CDCl$_3$, and the insets (a), and (b) show the shape of the sets of signals at highest and lowest chemical shift at −30°C.]

P2.10 Assuming that the spectrum of Fig. 2.43 is that of the pure compound referred to above, suggest an explanation for the fact that there are more peaks than might be expected (on a simple interpretation) in both the ring-carbon and the SiMe$_3$ regions.

P2.11 When an aqueous solution of [PtCl$_3$(dmso)]$^-$ was treated with glycine enriched in ^{15}N (H$_2$^{15}NCH$_2$CO$_2$H=Hgly), the ^{195}Pt NMR spectrum of the solution showed two signals, *1* and *2* in the table below. After the addition of one molar equivalent of alkali, only signal *2* remained. When the solution was heated, this signal was replaced by *3*. The species responsible for these signals appear to correspond to the two isomers of [PtCl(gly)dmso)] and to [PtCl$_2$(dmso)(Hgly)]. Compare the data given below with those in Table 2.22 and assign the signals to the three complexes. Explain the basis of your assignments. [Note that dmso binds to platinum through the sulphur atom.]

Signal	1	2	3
$\delta(^{195}Pt)$/ppm	3110	2747	2902
$^1J(Pt-N)$/Hz	244	226	330

P2.12 Treatment of a liquid-ammonia solution containing the anion $[V(CO)_6]^-$ with sodium metal gave an orange solution the ^{51}V NMR spectrum of which showed a single peak at -1962 ppm (relative to pure $VOCl_3$). After the addition of less than one molar equivalent of ethanol, the spectrum shown in Fig. 2.44 was obtained. What species are likely to be be present?
[Hint: the ^1H spectrum of this solution is shown in Fig. 2.15.]

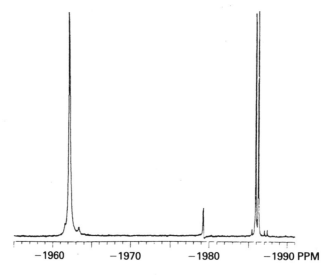

-1960 -1970 -1980 -1990 PPM

Fig. 2.44 — (Reproduced with permission from G. P. Warnock and J. E. Ellis, *J. Amer. Chem. Soc.*, **106** (1984) 5017.)

P2.13 When a solution of the complex *trans*-Pt(H)(CN)(PEt$_3$)$_2$ (*G*) was treated with HCl gas at $-90°C$, a new species (*H*) was formed which gave the ^1H and ^{31}P NMR data shown below (the chemical shift for ^{195}Pt was found by heteronuclear double resonance). On warming the solution, a gas was evolved and the spectra changed again (species *I*). What types of reactions are occurring? Use the data to identify the species present at the lower and higher temperature, and derive their stereochemistry.

	T °C	$\delta(^1H)$ ppm	$\delta(^{31}P)$ ppm	$\delta(^{195}Pt)$ ppm	$^1J(H-Pt)$ Hz	$^2J(H-P)$ Hz	$^1J(P-Pt)$ Hz
G		−7.75	19.21	373.7	827	15.5	2500
H	−90	−12.31	8.25	265.4	706	8.0	1677
		−18.40			1225	6.5	
I	0		16.41	373.3			2250

P2.14 Use the NMR data to identify the various rhenium complexes described below.

When *trans*-ReOCl₃(PPh₃)₂ was heated in benzene with the phosphine PBu$_2^t$Me, a dark green microcrystalline solid (*J*) was produced. The ¹H NMR spectrum contained two triplet resonances centered at δ 1.86 and 1.48 with apparent coupling constants of 7.3 and 12.9 Hz and integrated intensities in ratio 1:6. Analytical data confirmed that the expected phosphine-exchange reaction had occurred.

When *J* was treated with NaBH₄ in benzene in the presence of a phase-transfer catalyst, a brown oil was eventually isolated which could be neither crystallized nor purified further. NMR spectra showed that it contained about 80% of one component, *K*, which was thought to be a hydride complex, and which had the following ¹H parameters: δ1.57 ('triplet', 7.2 Hz, rel. intensity 6), 1.24 ('triplet', 12.9 Hz, rel. int. 18), −6.02 (triplet, 19.0 Hz, rel. int. 7). The ³¹P-{¹H} spectrum showed a singlet at 19.0 ppm; when the alkyl protons were selectively decoupled, the ³¹P resonance became an octet with lines in area ratio 1.2:6.7:20.6:35.0:21.4:7.3:1.2. [Hint: compare this with the possible theoretical ratios for coupling to various numbers of protons.]

When the PPr$_3^i$-analogue of *K* was heated in benzene with Ph₂PCH₂PPh₂ (dppm), the ³¹P-{¹H} spectrum of the solution (containing complex *L*) showed three resonances, at 51.0 (rel int. 2), 17.8 (1) and −20.3 (1) ppm. The position of the last of these is very similar to that of free dppm, and showed doublet structure with a coupling constant of 37 ppm. The 51.0 ppm signal was also a doublet, with separation 7 Hz, and the central signal was a doublet of doublets (7 and 37 Hz). The ¹H hydride resonance showed coupling to three ³¹P nuclei, two with ²J(P–H)=20.1 Hz and the third 15.4 Hz.

P2.15 Identify the species whose ¹⁹F NMR spectrum is shown in Fig. 2.45. It was

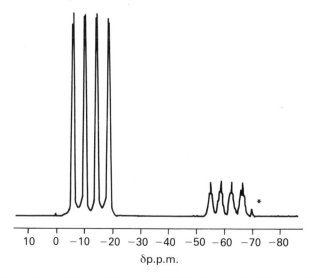

Fig. 2.45 — (Reproduced with permission from M. F. A. Dove, J. C. P. Sanders, E. L. Jones, and M. J. Parkin, *J. Chem. Soc., Chem. Commun.*, (1984) 1578.)

obtained by the oxidation of AsF_3 in the presence of Cl^-. It is thought to be of the type $AsF_nCl_{6-n}^-$ (note that the rightmost peak, marked with an asterisk, is a minor component which may be ignored). [^{19}F, 100%, $I=1/2$; ^{75}As, 100%, $I=3/2$; splitting due to Cl isotopes is not normally seen.]

P2.16   Fig. 2.46 shows part of the ^{19}F NMR spectrum of a $1:1$ mixture of UF_6 and

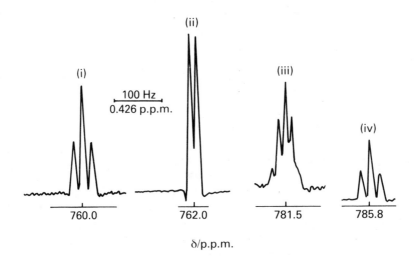

Fig. 2.46 — (Reproduced with permission from A. J. Downs and C. J. Gardner, *J. Chem. Soc., Dalton Trans.*, (1984) 2127.)

Me_3SiCl. The four sets of signals are believed to be due to two species formed by halogen-exchange. Suggest identities for the two species. [Chemical shifts are relative to CCl_3F.]

The full spectrum is listed below. Assign as many as possible of the signals to likely species, given that the chemical shift of UF_6 is 764.0 ppm and that, when Me_3SiCl is present in large excess, the major signal is at 774.3 ppm and small signals are seen at 781.0 and 746.1 ppm. All the spectra also contain a resonance at -159 ppm, due to Me_3SiF.

$\delta(^{19}F)$/ppm	786.45(s)	785.8(t)	782.6(t)	781.5(qnt)	781.0(s)
$^2J(F–F)$/Hz		18.8	24.0	13.3	
$\delta(^{19}F)$/ppm	774.3(s)	764.0(s)	762.0(d)	760.0(t)	755.5(s)
$^2J(F–F)$/Hz			13.3	18.8	
$\delta(^{19}F)$/ppm	753.0(d)	746.1(s)			
$^2J(F–F)$/Hz	24.0				

[s=singlet, d=doublet, t=triplet, qnt=quintet.]

P2.17 Derive the coupling pattern to be expected for a single proton coupled (a) to one ^{11}B and (b) to two equivalent ^{11}B nuclei ($I=3/2$).

P2.18 Reaction of B_3H_7.thf with the ylide Ph_3PCH_2 (prepared by reaction of $[Ph_3PCH_3]Br$ with sodium) gave a product with elemental analysis corresponding to the adduct $Ph_3PCH_2B_3H_7$. The ^{11}B-$\{^1H\}$ NMR spectrum showed resonances at -16.8 and -34.5 ppm (relative to $BF_3.OEt_2$) with area ratio 2:1. The ^1H NMR spectrum is shown in Fig. 2.47. Spectrum (a) is the result of broad-band ^{11}B

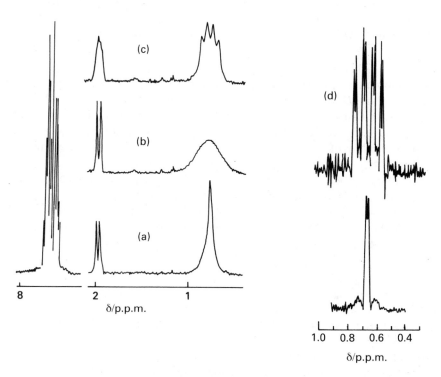

Fig. 2.47 — Reproduced with permission from P. Chung Choi and J. H. Morris, *J. Chem. Soc., Dalton Trans.*, (1984) 2119.

decoupling while in (b) and (c) the decoupling is restricted to the -34.5 and the -16.8 ^{11}B resonances respectively. Spectrum (d) is the same as (c) but enhanced by line-narrowing. Assuming a ring-structure for the B_3H_7-unit:

(i)   assign the major resonances in the ^1H spectrum;
(ii) account for the doublet structure of the δ 1.97 signal in (a) and (b), and its complexity in (c);
(iii) account for the shape of spectrum (d).

ANSWERS

A2.1 The shape of the spectrum is typical of that of π-allyl complexes; this is confirmed by comparison with the chemical-shift ranges in Tables 2.4 and 2.5.

A2.2 Select chemical shifts in accordance with Table 2.10. The coupling constants need to be translated into ppm on the basis that 100 Hz becomes 1.00 ppm.

Sketches are shown in Fig. 2.48, where the markers on the chemical shift scale are at -10, -15, and -20 ppm (left to right) except for the very last spectrum, where they are -5, -10, -15 ppm. Note the cases in which the hydride ligands are equivalent or not. Also note that for one isomer of (b), (d) and (f) the hydride ligands are magnetically non-equivalent even though related by symmetry; this is only of significance in (f) which therefore gives a second-order spectrum.

A2.3 (a) D (790 Hz), (b) A, (c) B (700 Hz), (d) C.

The platinum complexes are readily recognized by the presence of satellite resonances: B and D. In (a) the phosphine ligands are equivalent, and would split the hydride signal into a triplet; this must therefore have spectrum D. Complex (b) must necessarily have *cis* geometry; therefore the phosphines are non-equivalent and will give different values of ^2J(H–P), i.e. a doublet of doublets pattern, which must be A. Therefore, C corresponds to (d), confirmed by the simple triplet from equivalent *trans* phosphines. Spectrum B is basically a doublet of triplets, in which the doublet splitting corresponds to *trans* H–Pt–P. It must therefore be from complex (c).

Since the spectra have been run at 100 MHz, 1 ppm corresponds to 100 Hz. ^1J(H–Pt) is the separation between the satellites. Note that in B there is only one value of ^1J(H–Pt).

A2.4 *mer*, PMe$_2$Ph.

The methyl signals are a doublet and a triplet in intensity ratio 1:2, showing that two phosphine ligands are *trans*. The complex therefore has *mer* geometry. Since the hydride signal is a doublet of triplets, the hydride is *trans* to one phosphine and *cis* to the two others.

A2.5 $E=trans$-PtCl$_2$(PEt$_3$)$_2$, $F=trans$-Pt(H)Cl(PEt$_3$)$_2$.

Product E contains equivalent phosphines (δ $ca+12$ ppm) with ^1J(P–Pt)=(73 ppm)(32.4 Hz/ppm)=2370 Hz. The latter value is consistent with P *trans* to P, and suggests that a bridge-splitting reaction has occurred to give *trans*-PtCl$_2$(PEt$_3$)$_2$. This is the product expected on the basis of the *trans*-effect.

Product F is a hydride, $\delta(^1$H$)$ -16.9 ppm, in which the unique proton is coupled to two equivalent ^{31}P nuclei (triplet structure plus satellites). The line separation in the triplets gives the coupling constant ^2J(H–P)=14.5 Hz, of the correct magnitude for *cis* coupling. The ^{31}P spectrum confirms that the phosphine ligands are mutually *trans* [single resonance, δ ca 16 ppm, ^1J(P–Pt) 2750 Hz]. The value of ^1J(H–Pt) is given by the frequency separation of any pair of corresponding lines, e.g. 2332.0–1057.0 Hz=1275 Hz. F is therefore *trans*-Pt(H)Cl(PEt$_3$)$_2$.

A2.6 MCl$_3$L$_3$=2.XIX, (I)=2.XXI, (II)=2.XX.

The low-field spectra indicate the disposition of the neutral ligands. In every case,

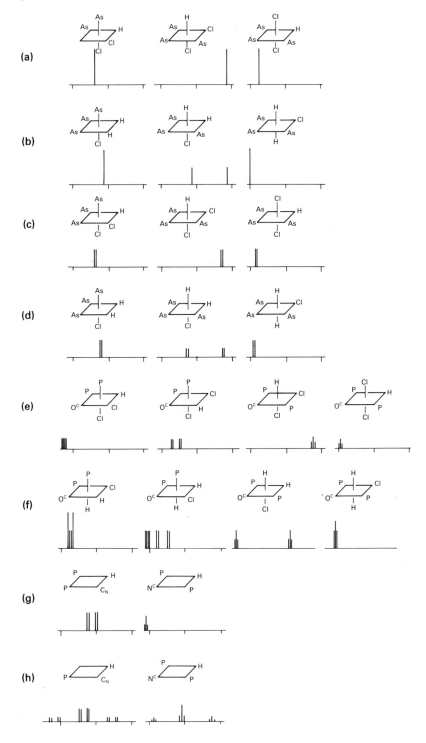

Fig. 2.48.

two sets of signals are seen in 2:1 intensity ratio, indicating a *mer* configuration. This is confirmed by the coupling (triplet : doublet) for the phosphines. The trichlorides therefore have structure **2.XIX**. In one case, isomer (II) of $RhHCl_2(PMe_3Ph)_3$, the methyl groups of the *trans* phosphines are not equivalent, indicating the lack of a plane of symmetry through these ligands, i.e. a *cis*-HCl_2 arrangement (**2.XX**).

The high-field spectra show the coupling of the hydride to ^{103}Rh and ^{31}P. The phosphine complexes are the most informative. For M=Ir, isomer (I) of $MHCl_2L_3$ shows a wide doublet of triplets, indicating that the hydride is *trans* to one ^{31}P and *cis* to two equivalent ^{31}P, structure **2.XX1**. The rhodium analogue gives essentially the same spectrum complicated by coupling to ^{103}Rh. Isomer (II) must therefore have structure **2.XX**, which is confirmed by the overlapping doublet-of-triplets shape of the hydride resonance.

2.XIX   2.XX   2.XXI

A2.7 See Fig. 2.49. The markers are at every 5 ppm in (e) and (g), and 20 ppm in (h). Spectra for complexes (f) will have the same shape as those for (e), although the absolute chemical shifts will differ. Note that for the first of complexes (e) the proton decoupling gives a simple spectrum even though the phosphine ligands are magnetically non-equivalent. For the *cis* complex in (h), all lines are actually doubled by P–P coupling, which is not resolved on this scale.

A2.8 (a) *cis*-$Pt(O_2SO_2)(PEt_3)_2$, (b) $Pt(OSO_3)(PEt_3)_3$.
(a) The most probable reaction is metathesis to produce *cis*-$Pt(O_2SO_2)(PEt_3)_2$ and AgCl; the neutral platinum complex is expected to be soluble in $CHCl_3$. The NMR data are consistent with this suggestion, the value of $^1J(P-Pt)$ being in the range for a *cis*-P_2Pt unit with electronegative ligands.
(b) In the presence of additional tertiary phosphine, the possibility of forming a tris-phosphine complex must be considered, with which the data are consistent. In a planar complex, two phosphine ligands are equivalent (*trans*) and couple with the third phosphine; 19.5 Hz is therefore the value of $^2J(P-P)$. The two values of $^1J(P-Pt)$ are consistent with P *trans* either to P or to a ligand with low *trans*-influence. The initial product is expected to be $Pt(OSO_3)(PEt_3)_3$. [The literature suggests that hydrolysis and/or exchange with $AgCl(PEt_3)$ may also occur — *J. Chem. Soc., Dalton Trans.* (1984) 2249.]

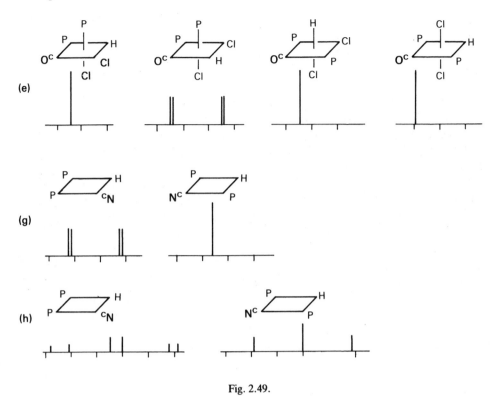

Fig. 2.49.

A2.9 Yes.

The spectrum does correspond to the expected product. The carbons of the cyclopentadienyl ring are at highest chemical shift, and its methyl has the second lowest chemical shift. The right-most peak (close doublet) corresponds to the CH_3 groups on silicon.

A2.10 Taking the cyclopentadienyl groups to be single ligands, the complex has pseudo-tetrahedral geometry about the metal. Since this arrangement possesses no plane of symmetry, neither the two rings nor the methyl groups of $CH(SiMe_3)_2$ form equivalent sets, and the resonances are multiplied.

A2.11 $1=2.XXII$, $2=2.XXIV$, $3=2.XXIII$.

The complex containing the Hgly ligand can only be present before alkali was added, and must be *1* or *2*. Since *1* disappears when alkali is added, it must be $PtCl_2(dmso)(Hgly)$ (in which the Hgly is N-bonded, **2.XXII**). Therefore *2* and *3* are the isomers of the chelated form $PtCl(gly)dmso$. The value of $^1J(Pt–N)$ for *3* suggests that N is *trans* to Cl (**2.XXIII**). Since $^1J(Pt–N)$ is similar for *1* and *2*, N must be *trans* to dmso in both **2.XXII**, **2.XXIV**), and the values are consistent with this. Configuration **2.XXII** is expected for *1* on the basis of the high *trans*-effect of dmso; chelation of the glycinate would lead directly to *2*:

Cf. J. Chem. Soc., Chem. Comm., (1977) 861.

2.XXII   2.XXIII   2.XXIV

A2.12 $V(CO)_5^{3-}$ and $HV(CO)_5^{2-}$.

In liquid ammonia, sodium is a strong reducing agent, while ethanol acts as an acid. Reduction of $V(CO)_6^-$ could give $V(CO)_5^{3-}$ (the singlet) which, on protonation would give the hydride $HV(CO)_5^{2-}$ (the doublet).

A2.13 H=2.XXVI, I=*trans*-PtCl(CN)(PEt$_3$)$_2$.

A likely initial reaction is oxidative addition of HCl to give a six-coordinate platinum(IV) complex, PtH$_2$Cl(CN)(PEt$_3$)$_2$. This is confirmed by the observation of two hydride resonances and the value of ^{1}J(P–Pt). The gas evolved on warming must be either H$_2$ or HCN, of which the former is more likely; this is confirmed by the fact that no hydride resonance is found for *I*. The reaction is therefore reductive elimination of dihydrogen, reforming platinum(II) as PtCl(CN)(PEt$_3$)$_2$. Four structures are possible for *H*, **2.XXV–2.XXVIII**. Since the two hydrides are not equivalent, all except **2.XXVI** can be eliminated. This is confirmed by the ^{31}P data which show that the phosphines are equivalent and therefore *trans*. The hydride with the lower value of ^{1}J(H–Pt) and the higher chemical shift is that *trans* to CN$^-$.

In *I* the phosphines are again equivalent, and ^{1}J(P–Pt) is low, showing that the geometry is *trans*.

Cf. *J. Chem. Soc., Dalton Trans.*, (1978) 877.

2.XXV   2.XXVI   2.XXVII   2.XXVIII

A2.14 J=*trans*-ReOCl$_3$(PBu$_2^t$Me)$_2$, K=*trans*-ReH$_7$(PBu$_2^t$Me)$_2$,
           L=ReH$_5$(PPr$_3^i$)$_2$(dppm-P).

Product *J* was expected to be *trans*-ReOCl$_3$(PBu$_2^t$Me)$_2$, which is confirmed by the two types of methyl group both showing 'virtual coupling' to the phosphorus nuclei. The ^{1}H spectrum of *K* similarly indicates the retention of the two phosphine ligands in *trans* positions. The partially proton-decoupled ^{31}P spectrum shows coupling only to the hydride ligands. Since the spectrum is symmetrical about its centre, the number of hydrides must be odd. An octet structure would derive from equal coupling to seven hydride ligands, which would give a theoretical intensity ratio of $1:7:21:35:35:21:7:1$, very close to that observed. The complex must therefore be *trans*-ReH$_7$(PBu$_2^t$Me)$_2$, which is fluxional, rendering all hydride ligands equivalent on the NMR time-scale. This formulation is consistent with the eighteen-electron rule.

Complex *L* contains some hydride ligands and has three phosphorus atoms coordinated to the metal. The dppm appears to be acting as a monodentate, having one donor atom coordinated and one free. The two PPr$_3^i$ ligands are therefore retained. In order not to exceed 18 electrons, a mole of hydrogen must have been lost: ReH$_5$(PPr$_3^i$)$_2$(dppm-P).
The data are taken from *J. Chem. Soc., Dalton Trans.*, (1990) 573.

A2.15 AsF$_5$Cl$^-$.
The spectrum consists basically of two quartets both of which show some fine structure. Quartet structure is due to ^{19}F bound to As. Each peak of the more intense quartet is split into a doublet, suggesting coupling to a single ^{19}F. There is therefore one unique fluorine atom. The second quartet is composed of symmetrical muliplets, indicating coupling to an odd number of ^{19}F nuclei, which must be three or five. The lines closest to the central line are relatively intense, and the whole does not have the appearance of a triplet; the intensities of the two major quartets are not in the ratio $2:1$. A *mer*=AsF$_3$Cl$_3^-$ configuration is therefore not possible. Quintet structure $(1:4:6:4:1)$ is credible, suggesting coupling to four ^{19}F nuclei; this and the relative intensities of the two quartets confirm the structure AsF$_5$Cl$^-$.

A2.16 (i)+(iv)=*cis*-UF$_4$Cl$_2$; (ii)+(iii)=UF$_5$Cl.
From the number of resonances, it is evident that all possible members and isomers of the series UF$_n$Cl$_{6-n}$ are present. The simplest species to identify are those which show F–F coupling. In any one structure, there are at most two sets of non-equivalent fluorine atoms, and both must show the same coupling constant. The splitting of the two signals for each species allow unambiguous assignment of the stereochemistry. For example, the two signals with ^{2}J(F–F)=18.8 Hz are both triplets; they therefore correspond to non-equivalent pairs of fluorine atoms, i.e. *cis*-UF$_4$Cl$_2$. Similarly, the 24.0-Hz coupled sets belong to *mer*-UF$_3$Cl$_3$, and the 13.3 Hz set to UF$_5$Cl. The major species in the presence of an excess of Me$_3$SiCl is likely to be UFCl$_5$, which therefore has δ=774.3, and the minor signals present are presumably *cis*- and *trans*-UF$_2$Cl$_4$. The remaining signals, at 786.4 and 781.0 ppm, are therefore due to *fac*-UF$_3$Cl$_3$ and *trans*-UF$_4$Cl$_2$.
Cf. *J. Chem. Soc., Dalton Trans.*, (1984) 2127.

A2.17 A coupling to one spin of 3/2 gives four equally intense lines separated by J(B–H). A second spin with the same coupling constant gives seven lines, separated by J(B–H), in intensity ratio $1:2:3:4:3:2:1$. Use the J-tree technique, when the splitting by the second ^{11}B gives lines which exactly overlap those of the first.

A2.18 (i) The major resonances at about δ 7.6, δ 1.97 and δ 0.76 correspond to phenyl, methylene, and borane-unit protons respectively. (ii) The CH_2-group is directly bonded to the phosphorus atom, and shows doublet structure due to coupling with ^{31}P. It must also be bonded to the unique boron atom (which resonates at -16.8 ppm) so that, when the other boron atoms are decoupled, as in (c), quartet structure is superposed on the ^{31}P splitting. (iii) Since there is only a single resonance for the seven protons of the B_3H_7 unit, these protons are undergoing rapid positional exchange. When coupling to two boron atoms is removed, only that to the unique boron atom remains, giving a quarter structure with each line doubled by coupling to ^{31}P.

3

Nuclear quadrupole resonance

Nuclear quadrupole resonance (NQR) spectroscopy lies between NMR and Möss-bauer spectroscopy. On the one hand, like NMR, it requires the application of radio-frequency radiation (but no magnetic field), on the other, like Mössbauer spectro-scopy, it depends on lack of symmetry in the electron density surrounding the nucleus. From the latter comes the possibility of obtaining information about structures and bonding.

The interaction between the nucleus and the surrounding electrons produces a series of energy levels, transitions between which can be induced by the application of electromagnetic radiation in the radio-frequency region. A very wide range of frequencies may be needed, 10^4–10^9 Hz, and, even though it is seldom necessary to scan the whole of this range, substantially wider fractions must be covered than for NMR. This makes accurate control difficult, and NQR is not widely practised and is regarded as somewhat 'exotic'. Although the type of information it gives can also be obtained by Mössbauer spectroscopy (Chapter 4), NQR is applicable to a different, wider range of nuclei (see Fig. 1.2, Table 3.1 and Appendix 2).

3.1 EXPERIMENTAL CONSIDERATIONS

Samples for NQR must always be solid. This is because the electric-field gradient which perturbs the nuclear energy levels (see below) is internal to the molecule under study. In liquids or gases, molecular tumbling occurs more rapidly than the characteristic frequencies used; the field gradient at the nucleus is then effectively averaged to zero. The inertia of the nuclei holds them more or less immobile, even though the molecule may be rotating. Gaseous or liquid samples may be studied by freezing, but it is necessary to ensure that the sample is crystalline: glassy solids often give very weak signals or none at all.

With all but the most modern Fourier-transform spectrometers, samples must be large (a few grams), and sample cells must be filled as efficiently as possible. For solids which melt near room temperature, it may be advantageous to pour the melt into the cell, taking care that crystallization occurs on cooling. Other solids usually

Table 3.1 — Common NQR isotopes

Isotope	Abundance/%	I	eQ/b	QCC$_0$/MHz[a]	
^{10}B	19.6	3	8.5×10^{-2}		
^{11}B	80.4	3/2	4.1×10^{-2}		
^{14}N	99.6	1	1×10^{-2}	ca	10
^{35}Cl	75.5	3/2	-1×10^{-1}		110
^{37}Cl	24.5	3/2	-7.9×10^{-2}		86.7
^{55}Mn	100	5/2	4×10^{-1}		
^{59}Co	100	7/2	3.8×10^{-1}		
^{63}Cu	69.1	3/2	-2.1×10^{-1}		
^{65}Cu	30.1	3/2	-2.0×10^{-1}		
^{75}As	100	3/2	2.9×10^{-1}	ca	150
^{79}Br	50.5	3/2	3.7×10^{-1}		770
^{81}Br	49.5	3/2	3.1×10^{-1}		645
^{121}Sb	57.3	5/2	-2.8×10^{-1}	ca	780
^{123}Sb	42.7	7/2	-3.6×10^{-1}	ca	1000
^{127}I	100	5/2	-7.9×10^{-1}	ca	2293

A more comprehensive tabulation can be found in Appendix 2.
[a] Absolute value of QCC corresponding to one p-hole.

require grinding to reduce the particle size and give good cell-packing, but this should be done as gently as possible to avoid producing crystal defects which can broaden and weaken the signals. Recrystallization to obtain small crystallites is better.

Spectra are frequently recorded with the sample at liquid-nitrogen temperature and, in principle a least, lowering the temperature gives an increase in signal intensity (see Chapter 1). If structural information is desired, it must be remembered that phase changes may occur on cooling; such changes can be detected by making measurements over a range of temperatures.

In order to aid the detection of weak signals, the spectra are usually presented in first- or second-derivative form, i.e. dI/dv or d^2I/dv^2 is plotted against v (see Fig. 3.1).

Unless the approximate positions of the lines are known, it will be necessary to make a preliminary scan to locate them and then to make a second sweep covering a narrow range with high resolution.

Compared with the frequency ranges normally scanned, the lines are very narrow; half-widths are usually in the range 1–10 kHz. This means that distinct signals may arise from atoms which differ only slightly in environment, e.g. through crystallographic non-equivalence. Thus, for SiCl$_4$ at 77 K, four ^{35}Cl resonances are observed, at 20.273, 20.408, 20.415, and 20.464 MHz, even though the molecules have tetrahedral symmetry at room temperature.

In cases where the sample contains more than one resonating nucleus, assignment of the signals may be complicated. The commonest case is for halogen-containing materials, since both chlorine and bromine have two isotopes. The

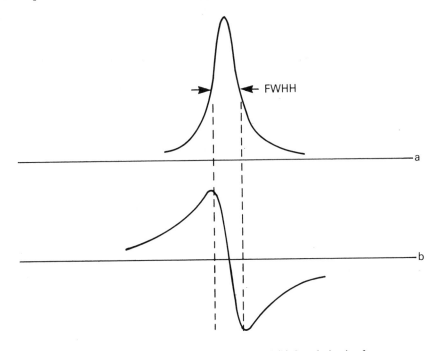

Fig. 3.1 — Spectra in (a) normal absorption and (b) first-derivative form.

isotopes of bromine have effectively equal abundances, so that all signals are doubled. The appropriate pairs can be recognized because the ratio of frequencies must be the same as the ratio of quadrupole moments, since the nuclei must be in identical environments. Similar considerations apply to the chlorine isotopes, except that they occur naturally in 3:1 ratio. Such problems of assignment are illustrated below (sections 3.3.1 and 3.4).

3.2 FUNDAMENTALS

The essential property which makes an isotope suitable for NQR is that its nucleus must have spin, I, greater than 1/2 (see Table 3.1 for common examples). Moreover, the nucleus must occupy a site in which the surrounding charge density has relativity low symmetry (lower than O_h or T_d). This is because nuclei with $I>1/2$ have non-spherical distributions of charge which, in the presence of an unsymmetrical electric field, allows only certain orientations of the nucleus. Each orientation corresponds to a different (quantized) energy, and transitions between these energy levels can be induced by radiation of suitable energy, usually in the MHz–GHz range.

The deviation from sphericity of the nucleus is measured by its quadrupole moment, eQ. The asymmetry of the electric field is measured by the electric-field gradient (EFG) in the z-direction (where z has been chosen as the direction of maximum gradient); it is the negative second derivative of the potential:

$$EFG = eq = V_{zz} = -\partial^2 V/\partial z^2$$

The interaction between these two is measured by their product, the quadrupole coupling constant, e^2qQ (QCC). (Note that in all these formulae, e stands for the absolute value of the electronic charge, i.e. a positive quantity; see below for a discussion of the sign of the EGF.) The various energy levels for a given nucleus are characterized by the magnetic quantum number, m_I, which has the values of $\pm I$, $\pm(I-1)$, etc., according to the formula

$$E_Q = \frac{e^2qQ[3m_I^2 - I(I+1)]}{4I(2I-1)} \tag{3.1}$$

Thus, when the spin is half-integral, $I + 1/2$ energy levels are obtained. Transitions are allowed between adjacent levels, the selection rule being $\Delta m_I = \pm 1$ (see Fig. 3.2); there are therefore $I - 1/2$ lines in the spectrum. For nuclei with integral spins, $I + 1$ levels and I transitions are possible. In either case, the magnitude of the QCC can be obtained from the line positions. Note that the QCC may be positive or negative. When it is negative, the energy-level pattern is reversed but the energies of the transitions are the same; the sign of the EFG cannot be determined from the spectrum.

The total spread of energies is very small, considerably less than kT (see Fig. 1.1), so that all levels have large populations. Transitions may therefore occur from any level, including those above the ground state. The intensity of the signal depends on the difference in population between the two levels, so that all lines have effectively the same intensity, which increases as the temperature is lowered.

3.2.1 Origin of the EFG

The electric-field gradient arises from an unsymmetrical distribution of electric charge about the nucleus. Such charge may come from non-bonding electrons (lone pairs, d-shell electrons), electrons in bonds, or charges on neighbouring atoms or ions.

The contribution of an individual charge to V_{zz} is given by $q(3\cos^2\theta - 1)/r^3$, where q is the size of the charge and θ and r its polar coordinates. The inverse cubic dependence on distance from the nucleus means that the valence electrons of the atom in question make a much larger contribution than shared electron-pairs in bonding orbitals which, in turn, have a greater effect than charges on surrounding atoms or ions. In the absence of polarization effects (Sternheimer effects, see section 3.3), the core electrons of the atom have spherical symmetry, and do not contribute directly. Thus, a lone pair of electrons always produces a substantial negative contribution to the EFG. When there are no lone pairs, quite large EFG values, positive or negative, may be found when the bond system has low symmetry.

There is some confusion in the literature about the definition and sign of the EFG and V_{zz}. This is partly due to the fact that the EFG is a tensor, of which

$$EFG = -\begin{vmatrix} V_{xx} & V_{xy} & V_{xz} \\ V_{yx} & V_{yy} & V_{yz} \\ V_{zx} & V_{zy} & V_{zz} \end{vmatrix}$$

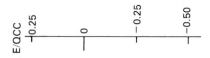

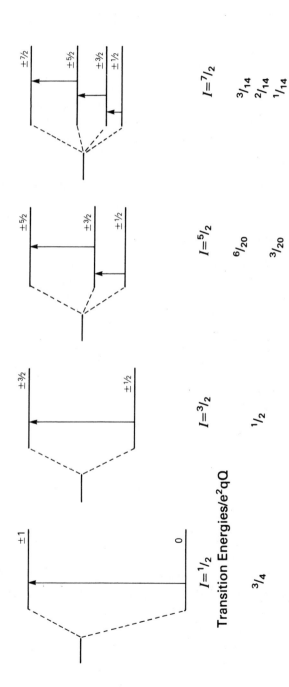

Fig. 3.2 — Energy levels and NQR transitions for various values of nuclear spin. The levels are labelled with the values of m_I.

only the principal component is normally considered. The negative sign comes from the definition of the electric field, which is the negative gradient of the potential, V. Thus, the EFG would be positive when V (and V_{zz}) was negative, i.e. when there was excess negative charge on the z-axis. However, common usage is to refer to either eq or V_{zz} as the EFG and to consider both as positive for a positive charge on the z-axis; this is the convention adopted here and in Chapter 4.

The fundamental requirement for the presence of an EFG is that the electron density in the z-direction be different from that in the x- and y-directions. Such differences are often referred to as electron imbalances, and imbalance in the populations of either the p- or the d-orbitals will give an EFG. The p-imbalance, U_p, may be defined as

$$U_p = N(p_z) - \tfrac{1}{2}N(p_y) - \tfrac{1}{2}N(p_x)$$

(The factor of $-\tfrac{1}{2}$ comes from the θ-dependence mentioned above, which makes charges on the z-axis twice as effective as, and opposite in sign to, those at right-angles to it.) A similar expression may be written for the d-imbalance:

$$U_d = N(d_{z^2}) - N(d_{x^2-y^2}) - N(d_{xy}) + \tfrac{1}{2}N(d_{xz}) + \tfrac{1}{2}N(d_{yz})$$

Such p- or d-imbalances will occur through bond-formation whenever the local symmetry of the atom is lower than O_h or T_d. Note that, when U_p or U_d is positive, there is excess electron density in the z-direction and, therefore, the EFG is negative (using the convention defined above).

For transition metal compounds, there may by a contribution to the EFG from the non-bonding d-electrons. This is often of opposite sign to that from the bonding electrons.

In the case where a single bond is made by the atom under investigation, it is often possible to find reference values for the QCC corresponding to particular electron configurations and thus to deduce the ionicity of the bond. This, the Townes–Dailey method, is discussed further below (see section 3.3.1). Where there are several bonds in a regular arrangement, involving two or more different ligands, the QCC may be sometimes be estimated by the partial-QCC method. Each ligand can be assigned a fixed contribution to the EFG which is weighted by the geometrical term, $3\cos^2\theta - 1$, where θ is the angle made by the ligand–atom bond to the z-axis. This method is applicable both to NQR and to Mössbauer data, and it is described in section 4.3.3.

3.2.2 The asymmetry parameter

The distribution of energy levels described above is only strictly correct for cases in which the electron density is the same in the x- and y-directions. This is commonly described as 'axial' symmetry: it occurs whenever the principal symmetry axis has order greater than two. If this is not the case, an additional parameter is required, the asymmetry parameter, η, given by

$$\eta = (V_{xx} - V_{yy})/V_{zz} \qquad\qquad (3.2)$$

The axes are normally chosen so that $|V_{zz}| > |V_{yy}| > |V_{xx}|$. Since V_{zz}, V_{yy} and V_{xx} sum to zero (Laplace rule), η can then never be greater than unity.

The effect of non-axial symmetry depends on the nuclear spin. In the case of half-integral spins, the energies are slightly affected. For $I = 3/2$, an exact expression can be found for the transition energy, which is $\frac{1}{2}e^2qQ(1 + \eta^2/3)^{1/2}$. In other half-integral cases, the energy of each level shows a complex dependence on η; for small values of η, the energy can be expressed as a power series in η^2 (see Table 3.2). In none of

Table 3.2 — Energy levels for non-zero η. $E_Q = e^2qQ(a + b\eta^2 + c\eta^4)$

I	$m_I{}^a$	a	b	c
1	+1	0.250	0.250	0
	±1	0.250	−0.250	0
	−0	−0.500	0	0
5/2	±5/2	0.250	0	0.0139
	±3/2	0.050	0	0.0750
	±1/2	0.020	0	0.0889
7/2	±7/2	0.250	0	0.0083
	±5/2	0.0357	0	0.0298
	±3/2	−0.1071	0	0.1107
	±1/2	−0.1786	0	−0.1488

^{a} Note that when $\eta > 0$, the m_I labels are not strictly correct, and also that 'forbidden lines' with $\Delta m_I = \pm 2$ may appear.
Cf. R. Bershohn, *J. Chem. Phys.*, **20** (1952) 1505.

these cases is the number of transitions affected, only their energies change slightly. For integral spins, however, the energies of the $\pm m_I$ substates become different. For example, for $I = 1$, the $m_I = \pm 1$ state splits as shown in Fig. 3.3, and two transitions are seen in place of one (it is now also possible for the third transition, with $\Delta m_I = 2$, to appear, although it has very low energy). (See Problem 3.7)

3.2.3 Effects of the magnetic field
The energies of nuclear spin states are affected by the presence of a magnetic field (this is the basis of NMR, see Chapter 2). Additional information can, in principle, be obtained by running an NQR spectrum with the sample in an applied magnetic field. A weak magnet will suffice, up to about 0.5 T being used.

The major effect of the magnetic field is to remove the degeneracy of the $\pm m_I$ states. Since the additional splitting is proportional to m_I and the selection rule remains unchanged, this would be expected to double the number of lines in the spectrum, and this is what often happens. However, for the $m_I = \pm 1/2$ states the two new states produced by the magnetic field are actually both mixtures of $m_I = +1/2$ and $m_I = -1/2$, so that transitions can occur from both of these levels to both of the $m_I = \pm 3/2$ levels (Fig. 3.4). In a magnetic field, therefore, the $m_I = 1/2 \rightarrow 3/2$

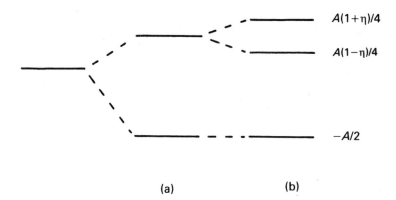

(a)   (b)

Fig. 3.3 — The effect of (a) an EFG and (b) non-axial symmetry on a nucleus with $I = 1$. The final energies are shown ($A = e^2qQ$).

transition splits into four. The intensities of the lines in a magnetically perturbed spectrum depend on the relative orientations of the EFG and magnetic-field axes.

Although unpaired electrons produce a magnetic field, no magnetic splitting is usually seen in the NQR spectrum of paramagnetic compounds. This is because the

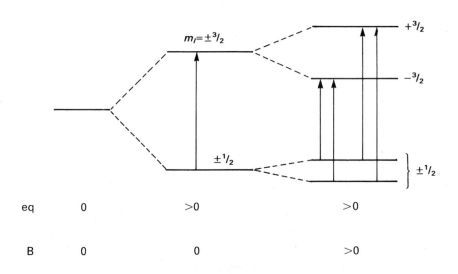

Fig. 3.4 — Effect of an external magnetic field for $I = 3/2$.

alignment of the spins on individual atoms is fluctuating rapidly and averages to zero on the NQR timescale.

3.3 INTERPRETATION OF THE SPECTRA

The EFG arises from asymmetry in the distribution of electrons about the nucleus. Filled shells of electrons can be regarded as spherical, to a first approximation. They can make a contribution to the EFG when they are polarized by the outer electrons or other charges (Sternheimer effect); this contribution can be substantial, but it is very difficult to quantify and is frequently ignored. As indicated above, non-bonding electrons in the valence shell of the atom being studied make the biggest contribution. Thus, atoms with one, two, or three lone pairs usually show quite large values of the QCC. The next largest contribution comes from the electrons in the bonds to the ligands. Since these are shared with the ligand atoms, their average distance from the NQR nucleus is greater than that of the non-bonding electrons. When the arrangement of bonds has symmetry less than cubic there will be a contribution to the EFG. For instance, there is an EFG at the nucleus of the M atom in a molecule MAB_5 which depends on the difference in the nature of the M–A and M–B bonds. The EFG increases as A and B become chemically more different. In a trigonal bipyramidal arrangement, MX_5, an EFG arises from the low-symmetry arrangement of five identical M–X bonds. These systems are treated in greater detail in section 4.3.3.

A few particular systems will now be discussed, to exemplify the interpretative methods used in NQR spectroscopy.

3.3.1   The halogens

The commonest isotopes investigated are those of chlorine and bromine, since the resonances occur in accessible regions, and there are many compounds available. Both elements have two naturally occurring isotopes, and all have $I = 3/2$. Thus, even when all the atoms occupy crystallographically identical sites, two resonances should be seen, one for each isotope. They may be recognized both from their relative intensities, 3:1 for $^{35}Cl{:}^{37}Cl$ or 1:1 for $^{79}Br{:}^{81}Br$, and from their relative frequencies, 1.27:1 or 1.19:1 respectively (this is the ratio of the nuclear quadrupole moments, since both must be in the same EFG). In the literature it is common for values for only one of the isotopes to be quoted; ^{37}Cl data are especially likely to be omitted because the low intensity of the signals makes their observation difficult (see Fig. 3.5).

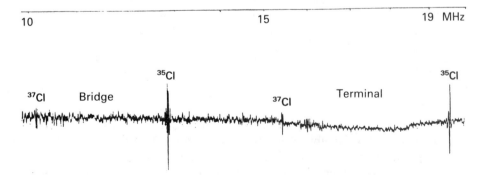

Fig. 3.5 — NQR spectrum of *trans*-[(Bu₃P)ClPd(μ − Cl)₂PdCl(PBu₃)] showing ^{35}Cl and ^{37}Cl resonances. (Reproduced with permission from C. W. Fryer and and J. A. S. Smith, *J. Chem. Soc. (A)*, (1970) 1030.)

In very many compounds, the halogen atom is covalently bound to only one other atom. There are thus three lone pairs and one bond pair of electrons. In this case, V_{zz} lies along the bond direction, and arises because the electron density in the p_z-orbital is smaller than that in p_x and p_y, which contain lone pairs. V_{zz} is therefore positive and increases as the electron density in the bond is drawn away from the halogen atom, i.e. as the other atom or group becomes more electronegative. This means that characteristic ranges of values may be recognized: ^{35}Cl–C, 31–40 MHz; ^{79}Br–C, 250–290 MHz; ^{35}Cl–Si, 15–20 MHz, etc. In the halide ions, however, there is a closed-shell configuration, all p-orbitals are equally populated, and the QCC is zero.

The value of the QCC may be used to estimate the effective electron configuration of the halogen atom in a compound, and hence the ionicity of the bonding. Two reference points are taken: (a) the free halogen molecule, in which the p_z-orbital of each atom effectively contains a single electron, and (b) the halide ion, which is spherical and must have zero EFG. The QCCs for the halogen molecules are very close to those determined for the free atoms in atomic-beam measurements, showing that the assumption of an s^2p^5 configuration for each atom in the molecule is not unreasonable. In considering other molecules, the observed QCC is used to interpolate between the two reference values:

$$\text{ionic character} = 1 - \text{QCC}/\text{QCC}_0 \tag{3.4}$$

where QCC_0 is the value corresponding to the halogen molecule (the halogen atom) (see Table 3.1). The value obtained represents the negative charge on the halogen. This is illustrated in Table 3.3 for interhalogen molecules. An indication of the reliability of the (rather crude) assumptions in this treatment can be gained by comparing the figures for the two different halogens in ICl.

Table 3.3 — Observed QCC-values and calculated bond ionicities (x) for interhalogens

	$e^2qQh^{-1}(^{35}\text{Cl})$/MHz	x		$e^2qQh^{-1}(^{127}\text{I})$/MHz	x
Cl–Cl	109	0	I–I	2293	0
Cl–F	146	+0.34			
Cl–Br	104	−0.95			
Cl–I	82.5	−0.24	I–Cl	2992	+0.30

This method of obtaining bond ionicities from NQR frequencies is a simplified version of a method originally developed by Townes and Dailey in the late 1940s (see Bibliography), and is widely used. It is, of course, not restricted to linear molecules. The ionicity of any element-halogen bond can be estimated in the same way. For example, for CCl_4, $\nu(^{35}\text{Cl}) = \frac{1}{2}e^2qQ = 40.63$ MHz. which gives an ionic character, $x = 0.37$. For $GeCl_4$ and $SnCl_4$ the frequencies are 25.66 $(x = 0.23)$ and 24.35 MHz

($x = 0.22$). These values are consistent with the decreasing electronegativity of the Group IV elements. (The value for $SiCl_4$, 20.90 MHz, appears to be out of line. This has been attributed to π-bonding involving donation of electron density from the p_x- and p_y-orbitals on chlorine into d-orbitals on silicon, which would have the effect of reducing V_{zz}.)

Similarly, transition-metal complexes can be compared. The data in Table 3.4 show an increase in covalency of the metal–halogen bond when the stereochemistry of nickel(II) changes from tetrahedral in the $[NiX_4]^{2-}$ anions and the Ph_3P complexes to square-planar in the other phosphine complexes. This presumably reflects the change in hydridization of the metal, from sp^3 to dsp^2, and the increase in covalency comes from the involvement of the d-orbitals; it also reflects the increased ligand-field which is present in the square-planar geometry. A more obvious cause for an increase in covalency comes from a change in oxidation state, as is seen in the data for the platinum(II) and platinum(IV) complexes and for gold(I) and gold(III) complexes. It is also noteworthy that the Pt–X bonds are more covalent than Ni–X, and this trend is repeated for Cu–X and Au–X.

Table 3.4 — ^{35}Cl and ^{79}Br NQR data and bond ionicities for some transition-metal complexes

Complex	e^2qQh^{-1}/MHz	x	Complex	e^2qQh^{-1}/MHz	x
$NiCl_2(PPh_3)_2$	18.94	0.83	$K_2[PtCl_4]$	35.90	0.67
$(Me_4N)_2[NiCl_4]$	17.90	0.84	$(Me_4N)_2[PtCl_4]$	35.74	0.66
$(Et_4N)_2[NiCl_4]$	18.41	0.83	$(pyH)_2[PtCl_4]$	37.01	0.66
$NiBr_2(PPh_3)_2$	146.22	0.81	$K_2[PtCl_6]$	51.63	0.53
$NiCl_2(PPr_3)_2$	31.70	0.71	$Na_2[PtCl_6].2H_2O$	52.83	0.48
$NiCl_2(PBu_3)_2$	31.98	0.71	$H_2[PtCl_6].6H_2O$	53.10	0.48
$NiBr_2(PPr_3)_2$	252.52	0.67			
$NiBr_2(PBu_3)_2$	253.06	0.67			
$Bu_4N[CuCl_2]$	19.30	0.81	$Bu_4N[AuCl_2]$	35.10	0.68
			$Na[AuCl_4].2H_2O$	55.20	0.50

Estimations of structures can be made from the numbers of lines observed and their intensities. For instance, in polyhalogeno compounds, it should be possible to observe distinct resonances from non-equivalent sites in the molecule. Thus, $C_6F_5PCl_4$ could have either structure **3.I** or structure **3.II**.

(R = C$_6$F$_5$)

3.I   3.II

Although the structures both have two geometrically distinct sets of chlorine atoms, the relative intensities of the two signals should be different: 3:1 for **3.I**, and 1:1 for **3.II**. In fact, four signals are seen for ^{35}Cl at 33.480, 34.380, 34.290, and 25.300 MHz (at 77 K). Since three of the frequencies are very similar and quite different from the fourth, structure **3.I** is likely. This case illustrates a common feature of NQR spectra: the line widths are relatively narrow, and separate signals are often seen from atoms which would be equivalent in the free molecule but are slightly non-equivalent in the crystal (see also the data for $SiCl_4$ given in section 3.1). The case of Me_3SbCl_2 is examined in section 3.4 below.

3.3.2 Group V elements

With the exception of phosphorus, all of the Group VA elements have isotopes suitable for NQR, although the typical frequency ranges are rather different. Data are readily available for ^{75}As and ^{121}Sb (the latter can also be obtained, with lower precision, from Mössbauer spectroscopy); ^{14}N resonates at very low frequencies (less than 2 MHz) and few data were available before the advent of double resonance methods.

The molecules R_3E (E = Group V element) have pyramidal structures with stereochemically active lone pairs. There is a substantial negative contribution to the EFG from the lone pair. This is offset by that from the E–R bonds, the contribution of which depends on the electronegativity of R and on the R–E–R bond angle (Table 3.5). When the lone pair is donated to a Lewis acid, its contribution to the EFG

Table 3.5 — Quadrupole coupling constants (MHz) for compounds of the Group V elements

Me_3N	3.161			$NF_3(g)$	7.07
Me_3As	96.7	Ph_3As	98.70	$AsCl_3$	158
Me_3Sb	630	Ph_3Sb	648	$SbCl_3$	384 (^{121}Sb)
		Ph_3Bi	669	$BiCl_3$	319
$Me_3As.AlMe_3$	82.4			Ph_3AsCl_2	118
$Me_3Sb.AlMe_3$	524			Ph_3SbCl_2	755
				Ph_3BiCl_2	1055

Some values are averages. The data for E = Sb are for ^{123}Sb, except for $SbCl_3$.

becomes less negative and the QCC diminishes by an amount depending on the extent of donation (A similar trend is seen for pyridine adducts, see Problem 3.7.) This is, in effect, a partial oxidation, and the arsenic(V) derivative Me_3AsBr_2 has a similar structure, $[M_3AsBr]Br$; the other R_3EX_2 derivatives appear to be trigonal pyramidal.

3.3.3 Transition metals

Several transition metals have suitable isotopes, but that most widely studied to date appears to be ^{59}Co. This is an $I = 7/2$ isotope with 100% natural abundance and an extensive chemistry. The carbonyl derivatives all have low-symmetry structures, often based on a trigonal bipyramid, which therefore show substantial QCC values (Table 3.6). Another example is $[Co\{P(OMe)_3\}_5]BPh_4$ which has $e^2qQh^{-1} =$ 153.7 MHz. The principal contribution to the EFG comes from the asymmetry in the

Table 3.6 — Quadrupole coupling constants (MHz) for ^{59}Co

$(OC)_4CoSnCl_3$	161.45	$[Co(NH_3)_5Cl]Cl_2$	31.74
$(OC)_4CoSnPh_3$	104.0	trans-$[Co(NH_3)_4Cl_2]Cl$	59.23
$Ph_3P(OC)_3CoSnPh_3$	114.5	trans-$[Co(en)_2Cl_2]Cl$	60.63
$Ph_3P(OC)_3CoPPh_3^+$	159.7	trans-$[Co(en)_2Cl_2]NO_3$	62.78
$Ph_3P(OC)_3Co-$			
$Co(CO)_3PPh_3$	146.8	cis- $[Co(en)_2Cl_2]NO_3$	33.71
en = $H_2NCH_2CH_2NH_2$			

distribution of the non-bonding 3d electrons. The metal atom has a d^8 configuration, in which all d-orbitals except d_{z^2} contain a pair of electrons. This will result in a large positive EFG, which is partly offset by donation from the ligands on the z-axis (the trigonal axis). Thus, the QCC for $Co(CO)_4SnX_3$ decreases markedly when x is changed from Cl to Ph.

Cobalt(III) forms many six-coordinate complexes. In this case, the metal has a low-spin d^6 configuration in which the d_{xy}, d_{xz} and d_{yx} orbitals are fully occupied (t_{2g}^6); the non-bonding electrons therefore make no contribution to the EFG. When the six ligands are identical, the complex has O_h symmetry and shows no NQR spectrum. Substitution of one or more ligands will produce an EFG the magnitude of which reflects the electronic difference in the two cobalt-ligand bonds. $[Co(NH_3)_5Cl]^{2+}$ therefore shows a modest QCC value (Table 3.6), and that for trans-$[Co(NH_3)_4Cl_2]^+$ is approximately twice as large. However, there is also a dependence on geometry, as is shown by the comparison of cis and trans $[Co(en)_2Cl_2]^+$. This is a system which responds well to the partial-QCC method (see section 4.3.3 and Fig. 4.6). Similar trends are shown by the isoelectronic manganese(I) carbonyl derivatives (Table 3.7).

Table 3.7 — Quadrupole coupling constants (MHz) for ^{55}Mn

$(OC)_5Mn–Mn(CO)_5$	3.28	$(OC)_5MnCl$	13.9
$(OC)_5Mn–Re(CO)_5$	8.67	$(OC)_5MnI$	19.8
$(OC)_5Mn–Co(CO)_4$	20	$(OC)_5MnSnPh_3$	18.3

3.4 COMPLEX SPECTRA

NQR spectra may be complicated for two possible reasons. First, the sample may contain two or more isotopes all of which give resonances in the same frequency range, e.g. ^{79}Br and ^{81}Br, or ^{35}Cl and ^{37}Cl, and ^{121}Sb and ^{123}Sb. Secondly, the crystal lattice may impose a lower symmetry than would be expected for a free molecule; in these circumstances, a separate resonance will usually be seen for each crystallographically distinct nucleus. These two possibilities will now be illustrated.

The compounds MCl_4 (M = Si, Ge, Sn) actually all give four ^{35}Cl resonance (and four ^{37}Cl resonances) separated by a few hundred kHz (see section 3.1). At room temperature, these materials are liquids, and contain tetrahedral molecules in which the chlorine atoms are equivalent. On solidification, slight differences are presumably introduced by crystal packing effects. The figures quoted above are average values.

Similarly, Na[AuCl₄].2H₂O is expected to contain a square-planar $AuCl_4^-$ anion in which the four chloride ligands are equivalent. In practice, four separate ^{35}Cl signals are seen, at 25.67, 27.46, 28.44 and 28.83 MHz (at 0°C). The chlorine atoms are not crystallographically equivalent, and one of them is hydrogen-bonded to the water molecules; this is presumed to be that with the lowest frequency.

When more than one NQR-active nucleus is present, signals from all should be seen. The data for Me_3SbCl_2 are shown in Table 3.8. In this case, one signal would

Table 3.8 — NQR frequencies (MHz) for R_3SbCl_2

R = CH₃	R = C₆H₅
11.07	12.80
14.05	16.25
60.30	56.75
99.30	90.02
120.6	107.2
180.9	162.2
198.6	177.3

be expected for each distinct ^{35}Cl or ^{37}Cl atom, two for each ^{121}Sb and three for each ^{123}Sb. Three molecular structures are possible (**3.III–3.V**), in one of which the chlorine atoms are non-equivalent. **3.III** and **3.IV** should also show non-zero values for the asymmetry parameter for the antimony isotopes. In the remaining structure, the chlorines are equivalent and the antimony lies on a three-fold axis, so that η should be zero. Assignments may most reliably be made on the basis of the frequency ratios, which should be: ^{35}Cl:^{37}Cl, 1.27:1.0, $^{121}Sb(5/2–3/2)$:$^{121}Sb(3/2–1/2)$, 2.0:1.0; $^{123}Sb(7/2–5/2)$:$^{123}Sb(5/2–3/2)$:$^{123}Sb(3/2–1/2)$, 3.0:2.0:1.0. (The ratios for the antimony isotopes are given for η = 0; ratios for other values of η can be calculated from Table 3.2.) Inspection then suggests that lines 1 and 2 are from ^{35}Cl and ^{37}Cl, lines 3, 5 and 6 from ^{123}Sb, and lines 4 and 7 from ^{121}Sb. Since all lines can be assigned, it is unlikely that any have been missed by being out of the range scanned, too weak, etc. It therefore seems likely that all the chlorine atoms are equivalent, suggesting structure **3.V**. Final confirmation is obtained from the frequency ratios for the antimony isotopes, which show that η(Sb) is zero.

    3.III   3.IV   3.V

Further examples of interpretation, including other isotopes, are given in Problems 3.1–3.7. It should also be noted that exactly the same quadrupole effects are seen in Mössbauer spectra, and further discussion is given in section 4.3.

BIBLIOGRAPHY

T. P. Das and E. L. Hahn, *Nuclear Quadrupole Resonance Spectroscopy*. Academic Press, 1958.
J. A. S. Smith, *J. Chem. Educ.*, **48** (1969) 39, A77, A147, and A243. 'Nuclear Quadrupole Resonance'. A useful series of articles dealing with all aspects of NQR.
T. B. Brill, *Adv. Nucl. Quad. Res.*, **3** (1978) 131. 'Nuclear Quadrupole Resonance as a Probe of Structure and Bonding in Organometallic Compounds'. Most useful — theory, experimental data, and interpretation.
M. Kubo and D. Nakamura, *Adv. Inorg. Chem. Radichem.*, **8** (1966) 257. 'Nuclear Quadrupole Resonance and its application in Inorganic Chemistry'. A short but useful review, giving a brief treatment of theory, interpretation and some data.
T. L. Brown, *Acc. Chem. Research*, **7** (1974) 408. 'Cobalt-59 Nuclear Quadrupole Resonance Spectroscopy'. Brief introduction to theory, and application to chemical systems involving cobalt.
E. Schemp and P. B. Bray, 'Nuclear Quadrupole Resonance Spectroscopy', in *Physical Chemistry, an Advanced Treatise*, Vol. 4, (D. Henderson, Ed.), Academic Press, 1970.
E. A. C. Lucken, *Nuclear Quadrupole Coupling Constants*, Academic Press, 1969. A compilation of data.
T. B. Brill, *J. Chem. Ed.*, **58** (1981) 519. 'Nuclear Quadrupole Resonance of R_3SbX_2 Compounds'. Short discussion of theory, experimental and interpretation.
C. H. Townes and B. P. Daily, *J. Chem. Phys.*, **17** (1949 782; **20** (1952) 35. 'Determination of Electronic Structures of Molecules from Nuclear Quadrupole Effects'. Two original papers with the theory of interpretation of QCCs in terms of orbital populations.

PROBLEMS

Answers are given on pp. 126–127

P3.1 Using the equation $E_Q = e^2qQ[3m_I^2 - I(I+1)][4I(2I-1)]^{-1}$ derive the energy-level diagram for nuclei with spins $I = 3/2, 5/2$ and $7/2$.

Using the selection rule $\Delta m_I = \pm 1$, show the possible NQR transitions.

P3.2 (a) Show by substitution into the E_Q formula why the highest energy level for a quadrupolar nucleus in a positive electric-field gradient (EFG) is always $e^2qQ/4$.

(b) What effect does changing the sign of the EFG from positive to negative have on the energy-levels?

(c) What effect does changing the sign of the EFG from positive to negative have on the spectrum?

P3.3 Explain the trend in the e^2qQ values for the ^{35}Cl nucleus in the following compounds:

Compound	H_3CCl	$MeCH_2Cl$	Me_2CHCl	Me_3CCl
e^2qQh^{-1}/MHz	68.06	65.96	64.14	62.12

P3.4 (a) Given the following data, calculate the charges on the bromine, carbon, and nickel atoms (explaining your reasoning):

Species	^{79}Br(atom)	CBr_4	$[NiBr_4]^{2-}$
e^2qQh^{-1}/MHz	769.8	530	129.2
		(estimated)	

(b) What do you expect the sign of the EFG to be at the bromine nuclei in these species? (Explain.)

P3.5 (a) Explain why the complex $[Co(NH_3)_6]^{3+}$ shows NQR resonance for ^{14}N but not for ^{59}Co (both isotopes have $I > 1$).

(b) Explain the observation that the ^{14}N quadrupole coupling constant is larger for the NH_3 molecule than for ammine complexes of transition metals.

P3.6 The cation $[Co\{P(OMe)_3\}_5]^+$ is isoelectronic and isostructural with $Fe(CO)_5$. Explain why it gives a three-line NQR spectrum. [Note: first explain why there is an EFG at the cobalt nucleus, then use the energy-level diagram derived above.]

$$[^{59}Co, 100\%, I = 7/2]$$

P3.7 Nitrogen-14 has $I = 1$ which gives, in the presence of an EFG, three states with energies $-A/2$, $+A(1-\eta)/4$, and $+A(1+\eta)/4$ (where $A = e^2qQ$).

(a) Calculate A and η for pyridine, given that signals are observed at 0.908, 2.984, and 3.892 MHz. [Note that, when $\eta \neq 0$, the strict selection rule $\Delta m_I = \pm 1$ no longer applies.]

(b) Compare your value for e^2qQ with those listed for the compounds below, comment on the observation of two sets of resonances in some cases, and interpret the data chemically.

Compound	$ZnCl_2py_2$	$ZnBr_2py_2$	$Zn(NCS)_2py_2$
A/MHz	2.95	2.89	2.89
η			0.92

Compound	$Zn(NO_3)_2py_2$	$pyHNO_3$	KNCS
A/MHz	2.66	1.09	2.43
η	0.57	0.69	

Data from G. V. Rubenacker and T. L. Brown, *Inorg. Chem.*, **19** (1980) 392.

ANSWERS

A3.1 The energies are the following multiples of the QCC;

(a) $+1/4$, $-1/4$
(b) $+1/4$, $-1/20$ (0.05), $-1/5$ (0.20)
(c) $+1/4$, $+1/28$ (0.0357), $-3/28$ (0.0107), $-5/28$ (0.179)

Transitions are possible between adjacent levels (cf. Fig. 3.2).

A3.2
(a) Substitute $m_I = I$.
(b) The order of energy levels is reversed.
(c) None.

A3.3 e^2qQ represents size of the 'hole' in the p-shell of Cl, which is progressively filled as the organic group becomes less electronegative, more electron donating.

A3.4 The QCC is 769.8 MHz for a bromine atom and zero for Br^-. Therefore, the charge on Br is $-[1 - QCC/(769.8\,MHz)]$

$$= -0.31 \text{ for } CBr_4, \quad -0.83 \text{ for } NiBr_4^-$$

Therefore charge on $C = +4 \times 0.31 = +1.24$

charge on Ni $= +4 \times 0.83 - 2 = +1.32$

(b) The sign of eq is governed by the difference in populations of the p_x, p_y and p_z orbitals. If the direction of the Br–X bond is chosen as z, p_z contains a bond pair, the other two contain lone pairs. There is thus less electron density along z, and the sign of the EFG is positive.

A3.5 (a) Cobalt(III) has a low-spin d^6 (t_{2g}^6) configuration, which gives equal electron density in the x, y, and z directions. The cation as a whole has exact octahedral (O_h) symmetry. There is therefore no EFG at the cobalt nucleus. On the other hand, the nitrogen atom in the ammonia ligand is attached to three hydrogen atoms and the cobalt atom and thus has an asymmetric environment (C_{3v} symmetry). The EFG reflects the difference between the N–H and N–Co bonds.

(b) In the NH_3 molecule, the EFG is largely due to the lone pair of electrons. When the lone pair becomes a bond pair, by donation to cobalt, its contribution to the EFG decreases because the electrons are now farther away from the nitrogen nucleus.

A3.6 The compound does not have cubic symmetry (D_{3h}), and cobalt(I) has a d^8 configuration. On both counts, the EFG is non-zero. Since $I = 7/2$, there are four energy levels, characterized by $m_I = \pm 7/2, \pm 5/2, \pm 3/2, \pm 1/2$. Since the selection rule is $\Delta m_I = \pm 1$, there are only three transitions possible.

A3.7 (a) The frequencies of the transitions are: $A.\eta/2$, $3A(1 - \eta/3)/4$, and $3A(1 + \eta/3)/4$. The sum of the last two is $3A/2 = 6.876\,MHz$. Hence, $A = 4.584\,MHz$, and $\eta = 2(0.980\,MHz)/(4.584\,MHz) = 0.40$. The magnitude of A represents primarily the presence of a lone pair of electrons on the nitrogen. The large value of η derives from the planarity of the ring, so that the components of the EFG are very different in the plane of the ring and perpendicular to it (lack of axial symmetry, N is at a site with C_{2v} symmetry).

(b) On coordination, the lone pair is partially delocalized onto the metal, giving a decrease in A. The acceptor ability of Zn^{2+} is affected to only a small extent by change in the other ligands, the difference being most marked for the nitrato complex (nitrate is a very weakly coordinated ligand). In the thiocyanate and nitrate, the second, low A-value must be due to the nitrogen atom in the nitrate or thiocyanate ligand. The value for nitrate changes little from that in the pyridinium salt, pyHNO$_3$, showing that bonding of one oxygen atom to Zn^{2+} has little effect on the nitrogen atom. There is a substantial change between ionic KNCS and the thiocyanate complex, indicating that the thiocyanate is bound through the nitrogen (an isothiocyanato complex).

4

Mössbauer spectroscopy

Mössbauer spectroscopy is more aptly described by its alternative title, nuclear gamma resonance spectroscopy and, in the Eastern bloc particularly, it is sometimes abbreviated to NGR. As this name implies, the nucleus is probed using gamma rays as the exciting radiation: a gamma-absorption spectrum is measured. The energy transitions concerned occur within the absorbing nucleus itself, but their magnitude depends on the density and arrangement of the extranuclear electrons. Thus, Mössbauer spectroscopy provides information about the type and spatial arrangement of the bonds made by the Mössbauer atom, its oxidation state and, in the case of transition metals, its spin state. As with the other nuclear techniques, only certain isotopes are suitable; those most commonly studied are listed in Table 4.1, together with some of their relevant properties.

Table 4.1 — Common isotopes for Mössbauer spectroscopy

Isotope	Natural abundance (%)	Spin states (gd)	(ex)	E_γ (keV)	Source isotope	Half-life	Line width^a (mm s^{-1})	Sign of $\Delta R/R$
^{57}Fe	2.2	1/2	3/2	14.4	^{57}Co	270 d	0.2	− ve
^{119}Sn	0.63	1/2	3/2	23.8	^{119m}Sn	240 d	0.8	+ ve
^{121}Sb	2.1	5/2	7/2	37.2	^{121m}Sb	77 y	1.8	− ve
^{127}I	100	5/2	7/2	57.6	^{127m}Te	105 d	2.0	− ve
^{129}I	0.6	7/2	5/2	27.7	^{129m}Te	33 d	0.8	+ ve
^{99}Ru	12.8	3/2	5/2	89.4	^{99}Rh	16 d	0.3	+ ve
^{193}Ir	61.5	1/2	3/2	73.0	^{193m}Os	30 h	1.4	+ ve
^{197}Au	100	3/2	1/2	77.3	^{197m}Pt	18 h	1.9	+ ve

^{a}Experimental line width in a 'good' spectrum.

As with most spectroscopic methods, an unconventional energy scale is used, and parameters are expressed in units of mm s^{-1}. This is because the exciting radiation is strictly monochromatic: gamma rays have precisely defined energies. The spectrum is therefore scanned by Doppler modulation; that is, the gamma source is mounted on a vibrator, and a controlled velocity is applied. The energy, E, perceived by the sample is then given by $E=E_\gamma(1+v/c)$, where v is the applied velocity and c is the velocity of light (i.e. of gamma rays). This in turn means that the energy scale is different for different isotopes, since the energy equivalent of 1 mm s^{-1} depends on the value of E_γ. This causes no problems, however, since it is seldom necessary to compare the energy scales for different isotopes. (Note also that, because the gamma energies are so precisely defined, it is essential to use the source appropriate to the isotope under study, in the same way that the correct frequency probe is selected for an NMR isotope.)

In practice, the spectrometer sweeps through a preset range of velocities (energies), and stores the gamma-counts transmitted by the sample in the channels of a multi-channel analyser, so that each channel corresponds to a known segment of the velocity range. The scan is repeated and the results are accumulated until a satisfactory signal:noise ratio is obtained. Depending on the strength of the gamma source, and the amount and concentration of the sample, this may take from a few minutes to many hours or even days. The normal statistical rule applies, i.e. that the signal:noise ratio is proportional to the square-root of the number of counts recorded. The peak positions are usually determined by computer analysis.

4.1 EXPERIMENTAL CONSIDERATIONS

4.1.1 The sample

For Mössbauer spectroscopy the sample *must* be a solid. Liquids, solutions or gases can be measured only by freezing them, although it is sometimes possible to obtain spectra from very viscous liquids or solutions. The reason for this restriction is that the nuclei under investigation must be prevented from recoiling when they absorb the gamma radiation (see below for a further explanation of this point). This means that the atoms containing the nuclei must be securely fixed in a solid lattice. In practice, only a small fraction of the nuclei do not suffer recoil, even in a solid, and it is this **recoil-free fraction** which gives rise to the observed spectrum. Clearly, the recoil-free fraction will be greatest for metals and alloys, and for ionic solids such as oxides, in which the lattice consists of a single framework throughout the whole crystal. In molecular materials, there is strong bonding only locally within each molecule; the recoil-free fractions are then low, and spectra are correspondingly less intense. For brevity, the recoil-free fraction is often referred to as the f-factor, and it is this property which largely governs the amount of sample required.

Samples are usually quite small, always less than 1 g (for a single substance), and sometimes as little as a few mg (Table 4.2). The amount needed depends both on the concentration of the Mössbauer element and on the f-factor. In addition to the points mentioned above, f also depends on the energy of the gamma rays. For isotopes with low gamma-energies such as ^{57}Fe (14.4 keV) and ^{119}Sn (23.8 keV), recoil effects are small and only small amounts of material are needed. As the gamma-energy rises so f falls, and larger samples have to be used; e.g. for ^{197}Au (77.3 keV), 200–300 mg of a molecular sample may be needed.

The recoil-free fraction can be increased by lowering the temperature, so that the majority of molecular materials require cooling to at least liquid-nitrogen temperature, and all samples involving isotopes with high excitation energies

Table 4.2 — Experimental considerations

Isotope	Amount needed[a] (mg/cm^2)	Temperature of measurement[b] (K)	Shape of spectrum	Other comments[c]
^{57}Fe	5–10	RT, LN	Doublet	Difficult in presence of heavy elements (self-absorption)
^{119}Sn	10–15	LN	Doublet	
^{121}Sb	10–15	LN, LH	Complex	Relatively poor resolution; difficult to obtain e^2qQ accurately when less than $ca \pm 10$ mm s^{-1} (300 MHz)
^{127}I	20–40	LN, LH	Complex	Most spectra poorly resolved; difficult to obtain e^2qQ accurately when less than $ca \pm 15$ mm s^{-1} (700 MHz)
^{129}I	15–20	LH	Complex	High cost of isotope for samples Radioactivity of samples Spectra well resolved Source activation needs high flux
^{99}Ru	50–100	LN, LH	Complex[d]	Source has relatively short lifetime, cannot be re-activated
^{193}Ir	50–250	LH	Doublet	Source activation needs high flux
^{197}Au	50–100	LH	Doublet	Short half-life (18 h) Source can be reactivated at medium flux

[a]Amount of the natural element, e.g. 5–10 mg of iron, containing 2.2% (100–200 μg) of ^{57}Fe.
[b]RT, room temperature; LN, liquid-nitrogen temperature (77 K); LH, liquid-helium temperature (4.2 K). The lower temperatures are required for molecular materials.
[c]Where source activation is mentioned, a source can be repeatedly used by re-activation in a nuclear reactor. Medium flux is ca 10^{12} n cm^{-1} s^{-1}, high flux $> 5 \times 10^{13}$ n cm^{-2} s^{-1}.
[d]The spectra approximate to doublets because the splitting of the 5/2 excited state is often less than the line width.

(i.e. $>ca$ 50 keV) must be measured at the temperature of liquid helium. In these latter cases, it is usually necessary to cool the source also.

The quantity of sample employed is usually quoted as the **sample thickness**, i.e. in terms of its mass per unit of cross-sectional area of the gamma-beam. However, the mass referred to is that of the *element* of interest, not that of the whole sample. Note also that no account is taken of the isotopic composition of the element, the total amount of the naturally occurring element present is quoted. Thus, satisfactory spectra for iron compounds can usually be obtained with sample thicknesses of 5–10 mg(Fe) cm^{-2}; such samples actually contain only 100–200 μg cm^{-2} of ^{57}Fe. In many cases, the beam cross-section is close to 1 cm^2, so that the thickness indicates the mass of the element required, from which the amount of sample may be calculated.

If the sample contains only a low concentration of the Mössbauer element, e.g. a dilute frozen solution, it would seem logical to increase the thickness of the absorber to give the desired mass in the beam. However, this is not always the best course, since increasing the thickness reduces the transparency of the absorber. This is particularly serious when the gamma-energy is low. The 14.4 keV radiation for ^{57}Fe is very easily absorbed, especially by the heavier elements (i.e. those of the first long Period and beyond). Under these circumstances, what is really desired is to obtain a satisfactory signal : noise ratio in the minimum time, and this is often better achieved with a thin absorber than a thick one. Although the thinner sample gives a smaller signal (absorption intensity), the improved transmission gives a higher rate of accumulation of data and a more rapid attainment of satisfactory statistics. For each sample there is an optimum thickness, and sometimes some experimentation is needed to find the best compromise.

When the Mössbauer isotope has relatively low natural abundance, very dilute systems can be studied by isotopic enrichment. For instance, many biological materials contain small numbers of iron atoms buried in proteins with molecular weights of tens or hundreds of thousands. To avoid impossibly long accumulation times, such materials are often specially prepared with iron containing 90% or more of ^{57}Fe.

The solid sample is normally ground to a fine powder and packed in a holder with windows transparent to the gamma rays. Care must be taken that the individual particles of the powder cannot move if adventitious vibrations occur. This is often assured by making a stiff mull with grease. When the crystallites are strongly anisotropic (e.g. plates or needles), the cell packing process can produce a partial orientation of the particles which may lead to asymmetry in the spectrum. This effect can be minimized by grinding the sample with an inert substance of cubic structure (and which does not contain the Mössbauer element). Suitable materials are sugar, alumina or boron nitride.

Solutions must be frozen rapidly, so that the solute does not have time to crystallize out. This is usually achieved by the use of a thin cell, preferably made of metal, which is immersed directly in liquid nitrogen to obtain a high rate of cooling. With aqueous solutions, it is usual to add glycerol or some other agent which encourages the formation of an amorphous glass. Of course, it is then necessary to ensure that the additive does not react with the sample.

4.1.2 Temperature of measurement

As indicated above, not all measurements can be made at room temperature. The biggest determining factor is the recoil-free fraction, which increases with decreasing temperature, decreases with increasing gamma-energy and with decreasing tightness of binding of the lattice. Even with the lowest-energy gamma rays, 14.4 keV for ^{57}Fe, many molecular or organometallic compounds do not give observable spectra at room temperature, but require cooling; the most convenient temperature is that of liquid nitrogen (77 K). It is even known for some ionic co-ordination compounds, which involve iron in both the cation and the anion, to show the spectrum of only one of the ions at room temperature. For ^{119}Sn (23.3 keV), only metallic and ionic samples can be measured at room temperatures; co-ordination compounds and organotin compounds must be run at liquid nitrogen temperature. With higher gamma energies, f is so small that liquid nitrogen is needed routinely and, in cases such as ^{197}Au (77.3 keV), liquid helium must be used.

4.1.3 The spectrometer

All spectrometers allow variation of the range of velocities covered, and many offer a choice of number of channels in which the spectrum may be accumulated (usually in multiples of 256).

Obviously, a large enough velocity range must be chosen to include all the peaks expected, and it is also necessary to include some baseline at the extreme velocities (this is needed to allow accurate computer fitting, since the baseline level is one of the parameters fitted). However, it is wasteful of data points to allow too much baseline, and the absorption peaks should be allowed to spread over about half the available data channels. Put another way, this means that the positions of the outermost peaks should be about three line widths from the first and last channels. This will sometimes mean making a preliminary wide scan to determine the extent of the spectrum, and then rerunning with a narrower range of velocities. Some spectrometers will allow the zero of velocity to be offset, which is useful in centralizing a spectrum in which all the peaks are on one side.

The number of channels devoted to a spectrum depends on the resolution required. Clearly, using more data points means better defined signals and enhanced resolution. It also means longer accumulation time to obtain the same signal : noise ratio, since the available counts are spread between a greater number of channels. In general, it is best to use the smallest number of channels unless precise determination of peak shapes is needed or the peaks are very narrow compared with the velocity range used (e.g. when using the six-line metallic-iron pattern to calibrate a wide velocity range).

4.2 FUNDAMENTALS

Gamma emission or absorption involves transitions between two different energy levels of the nucleus, the ground state and the first excited state. Since excitation involves a rearrangement of the sub-atomic particles in the nucleus, the total spin, I, is different for the two states (as are several other properties). Depending on the nature of the sample, any particular state may be split into as many as $2I+1$ sub-levels. The complexity of the spectrum is governed by the particular spin states of the nucleus and by the presence or absence of electric or magnetic fields. The basic shapes of the various types of spectra are now described, followed by an account of the significance of the major parameters.

4.2.1 The spectrum and its parameters

A Mössbauer spectrum is characterized by two major parameters, which are normally derived by using an appropriate computer program to make the best fit of a set of Lorentzian absorption peaks to the observed data. It is sometimes necessary to use a more sophisticated fitting program, e.g. when the sample is very thick or contains heavily overlapping subspectra.

The **isomer shift** governs the overall position of the spectrum on the energy (velocity) scale. It is measured as the position of the centre of gravity of the spectrum relative to that of the appropriate standard material. The standards normally used are shown in Table 4.3. The centre of gravity of the spectrum is calculated as the mean of the line positions, when the velocity of the each line is weighted by the theoretical line intensity. (Note that this is different from NMR, where the chemical shift for a set of signals in a coupling pattern is measured as the simple, unweighted average position of the lines.)

The second major parameter is the **quadrupole splitting** or **quadrupole coupling constant**. This has the same significance as for NQR spectroscopy (section 3.2)

Table 4.3 — Typical ranges of parameters for various oxidation states

Isotope	IS standard	Oxidation state	IS range (mm s^{-1})	QS(QCC) range (mm s^{-1})
^{57}Fe	Iron metal[a]	Iron(0)	-0.2 to -0.1	0.3 to 2.6
		Iron(II) —HS[b]	$+0.6$ to $+1.7$	1.0 to 4.5
		LS	-0.2 to $+0.4$	0.0 to 2.0
		Iron(III)—HS	$+0.1$ to $+0.5$	0.0 to 0.7
		—LS	-0.1 to $+0.5$	0.0 to 1.5
		Iron(IV)—HS	-0.2 to $+0.2$	0.0 to 1.0
		—LS	$+0.1$ to $+0.2$	1.5 to 2.5
^{119}Sn	SnO$_2$	Inorganic tin(II)	$+2.2$ to $+4.2$	0.5 to 2.0
		Inorganic tin(IV)	-0.5 to $+0.8$	0.0 to 1.0
		Organic tin(IV)	$+0.7$ to $+1.6$	1.5 to 5.5
^{121}Sb	InSb	Inorganic antimony(III)	-8 to -2	$(0.0$ to $+18)^c$
		Organic antimony(III)	-2 to 0	$(+15$ to $+18)$
		Inorganic antimony(V)	$+2$ to $+12$	$(0$ to $\pm 5)$
		Organic antimony(V)	0 to $+6$	$(0$ to $+30)$
^{127}I	(ZnTe)[d,e]	Iodine($-$I)	$+0.1$ to $+0.05$	$(0$ to $75)^{c,f}$
		Iodine(I)	-0.4 to -0.7	
		Iodine(III)	-1.0 to -1.2	$(+42$ to $+67)$
		Iodine(V)	-0.5 to -1.0	$(+19$ to $+27)$
		Iodine(VII)	$+0.8$ to $+1.5$	$(0$ to $+3)$
^{129}I	(ZnTe)[d,g]	Iodine(-1)	-0.5 to -0.2	$(0$ to $-105)^{c,f}$
		Iodine(I)	$+1.2$ to $+1.9$	
		Iodine(III)	$+2.8$ to $+3.5$	$(+60$ to $+95)$
		Iodine(V)	$+1.5$ to $+3.0$	$(+27$ to $+38)$
		Iodine(VII)	-2.3 to -4.5	$(0$ to $+5)$
^{99}Ru	Ru metal	Ruthenium(II)	-0.8 to -0.2	0 to 0.5
		Ruthenium(III)	-0.8 to -0.4	0.3 to 0.6
		Ruthenium(IV)	-0.3 to -0.2	0.2 to 0.6
^{193}Ir	Ir metal	Iridium(I)	-0.3 to $+0.3$	2 to 9
		Iridium(III)	-2.2 to $+0.4$	0 to 6
		Iridium(IV)	-1.3 to -0.9	0 to 4
^{197}Au	Au metal[h]	Gold(I)	-0.5 to $+7.0$	4 to 12
		Gold(III)	0.0 to $+7.0$	0 to 9

[a] Data are sometimes referred to Na$_2$[Fe(CN)$_5$(NO)].3H$_2$O. To correct these to the metallic-iron scale, subtract 0.26 mm s^{-1}.

[b] HS, high-spin; LS, low-spin.

[c] These values are more usually expressed in MHz; conversion factors are:
for ^{121}Sb, 1 mm s^{-1} = 30.00 MHz
for ^{127}I, 1 mm s^{-1} = 46.46 MHz
for ^{129}I, 1 mm s^{-1} = 32.58 MHz (on the ^{127}I scale, see text).

[d] ZnTe is the source material.

[e] Data are sometimes referred to CuI. To correct these to the ZnTe scale, subtract 0.14 mm s^{-1}.

[f] There is no clear distinction between iodides, covalent iodides, and iodine(I) compounds.

[g] Data are sometimes referred to CuI. To correct these to the ZnTe scale, add 0.41 mm s^{-1}.

[h] Data are sometimes referred to Au/Pt (the source). To convert to the Au scale, add 1.21 mm s^{-1}.

except that, in the Mössbauer case, two different nuclear states are concerned, each with its own spin, splitting pattern and coupling constant. This is made more clear below, but sections 3.2 and 3.3 should also be read.

In some cases, a **magnetic field** is present, either because the sample is magnetically ordered or because a field has been deliberately applied. This further complicates the shape of the spectrum, but analysis now allows the magnitude of the field at the nucleus to be calculated, which can sometimes provide important information.

The computer fitting usually gives the **width** of the absorption peaks. These should be close to the theoretical, or 'natural', value; acceptable values are shown in Table 4.1. If one or more peaks (or sets of peaks) show values appreciably greater than expected, it is likely that additional signals are present, and a new trial fitting should be made which includes a second subspectrum.

Finally, the spectrum will have a particular **intensity**. For complex spectra containing several lines, the intensities should be close to the theoretical ratio. However, if the sample is relatively thick, the stronger lines may show saturation effects, and a program which allows use of the 'thickness integral' should be applied. If the sample contains the Mössbauer atom in more than one distinct environment, or is a mixture of phases, two or more subspectra should be visible. The intensities of the subspectra will be related to the populations of the various sites. However, it should be noted that different sites may have different recoil-free fractions; in some cases the differences may be substantial, especially in molecular systems and when different oxidation states are involved. The relative intensities cannot then be used directly to determine the composition of the sample.

4.2.1.1 Simple spin states (1/2, 3/2)
Many of the isotopes of interest have spins of 1/2, 3/2 which, in the absence of a magnetic field, give a simple singlet or a symmetrical doublet spectrum (Fig. 4.1).

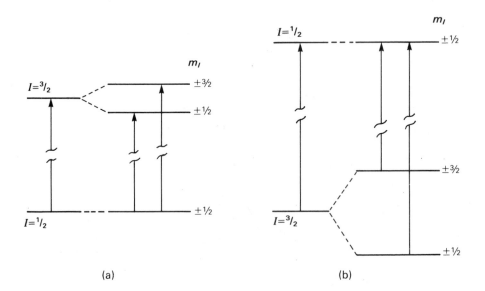

Fig. 4.1 — Mössbauer transitions for isotopes with spin states of 1/2 and 3/2, e.g. (a) ^{57}Fe, ^{119}Sn, (b) ^{197}Au.

The isomer shift, IS, is the position of the centre of the spectrum. The quadrupole splitting, QS, is the separation between the two absorption peaks of the doublet; it appears because the $I = 3/2$ state is split into two sublevels by the presence of an electric-field gradient (EFG, see section 3.2). The magnitude of the QS is one-half of the quadrupole coupling constant (QCC) for the $I = 3/2$ state. In a singlet spectrum the QS is zero or very small, and it is not possible to resolve two peaks which are closer than 0.4 times the line width. Although it is usual to make a computer fitting to the experimental points, quite good values can be derived by eye from a well-defined singlet or doublet spectrum. This is not usually the case for isotopes with higher spins. The doublet is sometimes asymmetric, and possible reasons for this are discussed in section 4.5.1.

4.2.1.2 Higher spin states

For $I > 3/2$, the presence of an EFG gives a splitting of the nuclear states into $I + 1$ or $I + 1/2$ sub-levels (depending on whether I is integral or half-integral, see section 3.2). The number of transitions between the ground and excited nuclear states is now governed by the selection rule $\Delta m_I = 0, \pm 1$. For example, several useful isotopes have spin states of 5/2, 7/2, so that the selection rule leads to eight transitions (Fig. 4.2).

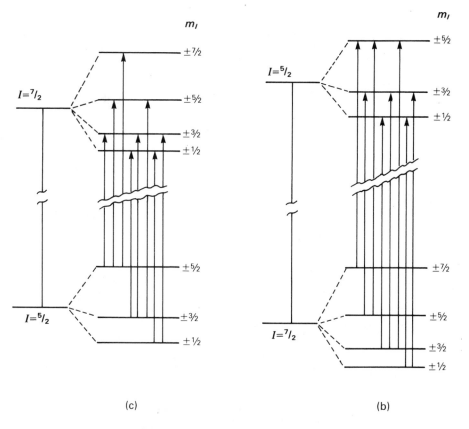

(c)   (b)

Fig. 4.2 — Transitions for isotopes with spin states of 5/2 and 7/2, e.g. (a) ^{121}Sb, ^{127}I, (b) ^{129}I.

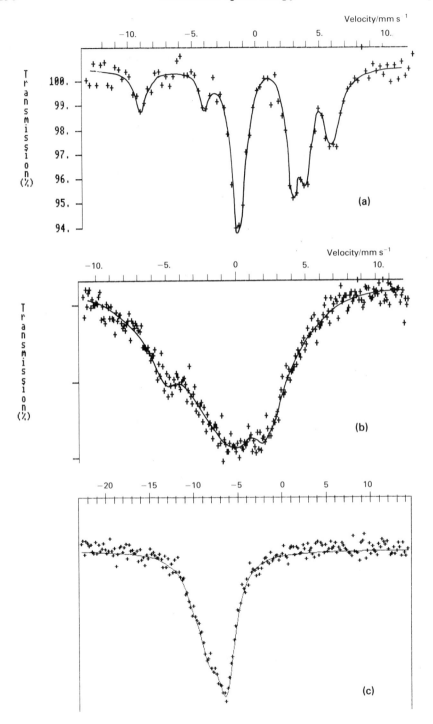

Fig. 4.3 — (a) ^{129}I Mössbauer spectrum of PtI$_2$(cod), (b) ^{127}I spectrum of PtI$_2$(cod), (c) ^{121}Sb Mössbauer spectrum of [Ag(Ph$_3$Sb)$_4$]BF$_4$. [cod = 1,5-cyclooctadiene.]

The line positions are functions of the magnitudes of the EFG and the nuclear quadrupole moments for the two states; since the EFG is common to both, and the ratio of the two quadrupole moments is usually known, the line positions can be given in terms of the quadrupole coupling constant (QCC) of the ground state (Table 4.4).

Table 4.4 — Positions and intensities of the absorption lines for 5/2, 7/2 isotopes

Transition	Relative intensity	Line position/mm s^{-1}		
		^{121}Sb (1.34)	^{127}I (0.896)	^{129}I (1.231)
5/2–3/2	0.0119	− 0.3936	− 0.3462	+ 0.4149
5/2–5/2	0.0714	− 0.2021	− 0.2180	+ 0.2720
5/2–7/2	0.2500	+ 0.0850	− 0.0255	+ 0.0578
3/2–1/2	0.0357	− 0.1893	− 0.1104	+ 0.1170
3/2–3/2	0.1191	− 0.0936	− 0.0462	+ 0.0456
3/2–5/2	0.1786	+ 0.0979	+ 0.0820	− 0.0972
1/2–1/2	0.2143	− 0.0393	+ 0.0397	− 0.0676
1/2–3/2	0.1191	+ 0.0564	+ 0.1038	− 0.1390

The data refer to QCC = + 1 mm s^{-1} and $\eta = 0$. The line positions for a real case are given by multiplying these coefficients by the QCC and adding the IS. The number in parentheses after each isotope is the ratio of the quadrupole moments, Q_{ex}/Q_{gd} .

The intensities of the various lines are not equal, and the spectrum may often have a very asymmetrical shape (see Fig. 4.3). The isomer shift is now not necessarily marked by the apparent centre of the spectrum. When the EFG has axial symmetry ($\eta = 0$, see section 3.2.2), the positions of the lines and their intensities can be given simple values (Table 4.4), and these can be included in a suitable fitting program. As η increases from zero, the $\Delta m_I > 1$ lines begin to appear, and the relative intensities of the other lines change; when η is appreciable (> ca 0.3) it is best to use a program which allows full definition and diagonalization of the Hamiltonians for the nuclear states. When η is close to unity, the spectrum becomes symmetrical about its centre.

4.2.1.3 Magnetic splitting

In some cases, even the simple (1/2, 3/2) spectra show additional splitting owing to the presence of a magnetic field. This is most often encountered for ^{57}Fe in ferromagnetic materials, but it may also be seen in paramagnetics and antiferromagnetics at which have been cooled to low temperatures and have become magnetically ordered. Alternatively, magnetic splitting may be deliberately induced by the application of a strong external magnetic field (> 5 T). The effect of the magnetic field is to remove the degeneracy of the $+ m_I$ and $− m_I$ substates, so that the number of substates is doubled. The selection rule is unchanged, so the the spectrum now consists of six lines (Fig. 4.4), as shown for iron metal and Fe_2O_3 in Fig. 4.5. The latter has a small quadrupole splitting superposed on the magnetic splitting. The positions of the lines as a function of the magnetic field and quadrupole splitting are given in Table 4.5. If the sample is composed of randomly oriented crystallites, the

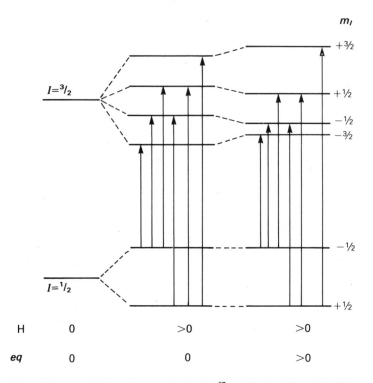

Fig. 4.4 — Zeeman splitting of energy levels for ^{57}Fe. The $I = 1/2$ and $I = 3/2$ states have magnetic moments which differ in sign.

intensities of the six lines are in the ratio $3:2:1:1:2:3$. Orientation of the crystallites can affect these ratios considerably. In particular, if the γ-ray direction is parallel to the magnetic field direction, the intensities of lines 2 and 5 become zero.

The application of a magnetic field provides a means of determining the sign of EFG for $I = 1/2$, $3/2$ isotopes, since the two peaks of the quadrupole doublet split in different patterns. For ^{57}Fe, the transitions to the $m_I = \pm 1/2$ level give a triplet, while those to the $m_I = \pm 3/2$ give a doublet (this may be seen by drawing the line positions shown in Table 4.5 for a case in which the quadrupole splitting is greater than the magnetic splitting. For other isotopes with these spin states, the lines may be too broad to allow full resolution of this pattern.

4.3   SIGNIFICANCE OF THE PARAMETERS

4.3.1   The isomer shift

The IS measures the total electron density on the Mössbauer atom. Strictly, it is the electron density *at the nucleus* which is important (relative to that in the standard material). The nucleus interacts with the electron density in a manner which raises the energy levels by an amount which is proportional both to the size of the nucleus and to the magnitude of the electron density. The effect arises because the nucleus

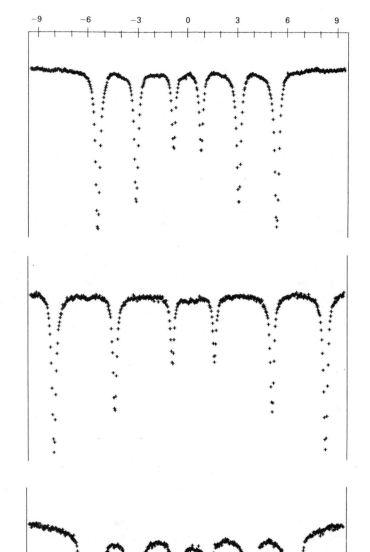

Fig. 4.5 — ^{57}Fe Mössbauer spectra for (a) metallic Fe, (b) Fe_2O_3, and (c) Fe_3O_4.

Table 4.5 — Positions of the six lines (mm s^{-1}) of a magnetically split ^{57}Fe Mössbauer
spectrum as a function of the QS^a and magnetic field (B)

Line 1: IS + 0.5000 QS − 0.1615 (B/T) mm s^{-1}
Line 2: IS − 0.5000 QS − 0.0934 (B/T) mm s^{-1}
Line 3: IS − 0.5000 QS − 0.0254 (B/T) mm s^{-1}
Line 4: IS − 0.5000 QS + 0.0254 (B/T) mm s^{-1}
Line 5: IS − 0.5000 QS + 0.0934 (B/T) mm s^{-1}
Line 6: IS + 0.5000 QS + 0.1615 (B/T) mm s^{-1}

^{a} The above relationships assume that the EFG axis is parallel to the magnetic field. For the general case,
0.5000 QS should be replaced by ε, where $\varepsilon = e^2qQ(3\cos^2\theta - 1)/8$ and θ is the angle between the EFG and
magnetic-field axes.

has different radii in the ground and excited states, so that interaction with the
electrons affects the two energies differently. When the electron density is altered
the transition energy changes, and this change is measured as an isomer shift between
the two systems. Since only s-electrons have finite probability of occurring at the
nucleus, the IS measures principally the difference in valence s-orbital populations of
the Mössbauer atom between the sample and the source.

$$IS \propto (\Delta R/R).\Delta|\psi(0)|^2$$

where R and ΔR are the nuclear radius and its change on excitation, and $\Delta|\psi(0)|^2$ is
the difference in electron density at the nucleus (which is usually taken as proportional to the valence-orbital population).

For isotopes for which R_{ex} is larger than R_{gd}, $\Delta R/R$ is positive (see Table 4.1),
and any factor which increases the s-orbital population increases the IS. Thus,
increase in the covalency of the bonds or in donation by the ligands increases the total
electron density and, therefore, the s-electron density. The presence of a lone pair
with high s-character gives a large positive IS. The IS is also increased, but to a much
smaller extent, by factors which decrease the p- or d-populations, owing to the
decrease in shielding of the nucleus from the s-electron density. The most obvious
such factor is an increase in oxidation state of a transition-metal atom. A decrease in
co-ordination number also increases the contribution of the s-orbital and decreases
the p- and d-populations, raising the IS, e.g. tetrahedral co-ordination, involving sp^3
hybridization, would give a higher IS than octahedral, d^2sp^3, hybridization. Note,
however, that for the most common isotope, ^{57}Fe, and for several others, $\Delta R/R$ is
negative and all these effects are reversed.

4.3.2 Quadrupole splitting

Quadrupole splitting arises from the presence of an EFG at the Mössbauer nucleus,
in exactly the same way as described for NQR spectroscopy ($q.v.$, section 3.2). Thus,
the magnitude of the QS or QCC represents the asymmetry of the electron cloud
around the nucleus. Any effect which gives a different population in the p_z orbital
from that in the (p_x, p_y) orbitals, or in the d_{z^2} from $d_{x^2-y^2}$, or in d_{xy} from (d_{xz}, d_{yz}), will

produce a splitting of the spectrum which increases as the population imbalance increases. Such imbalances may occur directly through the presence of lone pairs of electrons, i.e. non-bonding electrons, or indirectly through the bonding interactions. Owing to the inverse cubic dependence of the EFG on the distance of the charge from the nucleus, the effects of non-bonding electrons are usually bigger than those of the bonding interactions, since bonding electrons are shared between the Mössbauer atom and its ligands while non-bonding electrons are localized entirely on the Mössbauer atom. Hence, lone pairs of electrons make a large contribution to the EFG, e.g. in tin(II) or antimony(III) compounds. For transition-metal atoms, there is often an imbalance in the d-shell: in octahedral geometry only d^3, d^8, d^{10}, high-spin d^5 and low-spin d^6 configurations make no contribution to the QS. Asymmetry in the population of the bonding orbitals occurs when the ligands are not all equivalent, or the Mössbauer atom is at a site with symmetry less than O_h or T_d. The magnitude of the EFG then depends on the difference in covalency of the bonds. In transition-metal compounds, the contributions of the d-shell and the bond-asymmetry are of opposite sign and partially cancel. In ionic solids, small QS values may arise from low-symmetry arrangements of the ions in the lattice.

4.3.3   The additive model

For several systems it has been found possible to make reasonable estimates of the IS and QS (QCC) by summing the contributions from the individual ligands. This assumes that the bonding of any particular ligand is unaffected by the other ligands present, which is approximately true when the bonding is not highly covalent. It is also necessary that there be no contribution from non-bonding electrons.

Thus, the IS for any given set of ligands may be estimated simply by adding the appropriate values:

$$IS = \Sigma_i \langle L_i \rangle \tag{4.1}$$

A scale of partial-IS values, $\langle L \rangle$, is given in Table 4.6. Note that values quoted for bidentate ligands are those per donor-atom, e.g. the value for dppe, $\langle \frac{1}{2}dppe \rangle$, represents the contribution of only one phosphorus donor atom and the partial-IS value must be doubled to obtain the contribution of the whole ligand.

In a similar way, the QS may be estimated from the partial-QS values of the ligands, [L] (Table 4.7). The QS, however, depends not only on the identity of the ligands but also on their geometrical arrangement. The partial-QS value for each ligand must therefore be used in conjunction with the angle (θ) between the metal–ligand bond and the z-axis (usually the axis of highest symmetry):

$$QS = \Sigma_i [L_i](3 \cos^2\theta_i - 1) \tag{4.2}$$

Note that the sign of the QS given by eqn (4.2) is actually that of the EFG, i.e. what is calculated is $\frac{1}{2}e^2q|Q|$. Note also that different scales of [L]-values are required for different co-ordination numbers, since the hybridization, and hence the p- or d-character of the bonding orbital, changes with stereochemistry.

Table 4.6 — Partial-IS values, $\langle L\rangle$/mm s^{-1}, for various isotopes.

$$IS = \Sigma_i \langle L_i \rangle$$

Ligand	^{57}Fe^{II}(LS)	^{99}Ru^{II}	^{99}Ru^{III}	^{193}Ir^{III}	^{197}Au^{I}
Cl$^-$	+0.07	−0.10	−0.12	−0.36	+0.92
Br$^-$	+0.10	−0.12	−0.13	−0.37	+0.83
I$^-$	+0.10				+0.87
H$_2$O	+0.07		−0.12		
Me$_2$S					+1.72
NCS$^-$	+0.02		−0.08	−0.28	
NH$_3$	+0.04	−0.15	−0.08	−0.25	
py		−0.12		−0.13	+1.60
PPh$_3$	+0.04				+2.59
PMe$_2$Ph				+0.10	+2.61
½dppe^a	+0.04			−0.02	
CN$^-$	−0.02	−0.04		+0.04	+2.17
H$^-$	−0.11			+0.33	
CO	<−0.04	+0.22		ca +0.6	+2.10
NO$^+$	−0.23	ca +0.3			

^{a} Note that for bidentate ligands, the value given refers only to one donor atom.

Table 4.7 — Partial-QS values, [L$_i$]/mm s^{-1}, for various isotopes.

$$QS = \Sigma_i [L_i] (3 \cos^2 \theta_i - 1)$$

Ligand	^{57}Fe^{II} ^{b}oct.	^{99}Ru^{II} oct.	^{193}Ir^{III} oct.	^{197}Au^{I} lin.	^{119}Sn^{IV} tet.	tba.	tbe.	oct.	^{121}Sb^{V} ^{a} oct.
Cl$^-$	−0.30	0.00	0.00	−1.57	0.00	0.00	0.00	0.00	0.00
Br$^-$	−0.28	+0.06		−1.59	−0.07	0.00		0.00	0.00
I$^-$	−0.29		+0.33	−1.44	−0.17	−0.08		−0.14	−0.46
H$_2$O	−0.45					+0.18			
Me$_2$S				−1.89					
NCS$^-$	−0.51	−0.02			+0.21	+0.065		+0.07	+0.23
NH$_3$	−0.52	−0.08	−1.03						
py	−0.47		−0.75	−1.83		−0.035		−0.10	−0.33
PPh$_3$	−0.53			−2.38					
PMe$_2$Ph	−0.62		−1.46	−2.41					
½dppe	−0.65								
CN$^-$	−0.84	−0.10		−2.50					
H$^-$	−1.04		−2.24		−1.06				
CO	−0.55	ca −0.02		−1.90					
NO$^+$	+0.01	+0.10							
C$_6$H$_5^-$					−1.26	−0.89	−0.93	−0.95	−3.09
CH$_3^-$					−1.37	−0.94	−1.13	−1.03	−3.35

Note that for bidentate ligands, the values given refer to a single donor atom.
Note also that when these values are inserted into the above equation, the sign obtained is that of the EFG.
^{a} Data for ^{121}Sb are partial QCC values
^{b} oct., octahedral; lin., linear; tet., tetrahedral; tba, trigonal-bipyramidal, axial; tbe, trigonal-bipyramidal, equatorial.

Thus, for *trans*-[FeCl$_2$(dppe)$_2$]:

$$IS = 2\langle Cl^- \rangle + 4\langle \tfrac{1}{2}dppe \rangle$$
$$= 2(0.07 \text{ mm s}^{-1}) + 4(0.04 \text{ mm s}^{-1})$$
$$= 0.30 \text{ mm s}^{-1}.$$
$$QS = [Cl^-](3 \cos^2 (0) - 1) + [Cl^-](3 \cos^2 (180°) - 1)$$
$$+ 4[\tfrac{1}{2}dppe](3 \cos^2 (90°) - 1)$$
$$= (-0.30 \text{ mm s}^{-1})(2) + (-0.30 \text{ mm s}^{-1})(2)$$
$$+ (-0.65 \text{ mm s}^{-1})(-4)$$
$$= 1.20 \text{ mm s}^{-1}.$$

These values compare favourably with those observed: 0.35 mm s^{-1} and 1.36 mm s^{-1}. (Satisfactory agreement is usually considered to be within 0.07 mm s^{-1} for the IS and 0.4 mm s^{-1} for the QS.)

The partial-IS and partial-QS values for a given ligand may be interpreted as measures of the amount of electron density transferred by the ligand to the *s*- and *p*-components of the bonding orbital of the Mössbauer atom.

For systems involving the same ligands in different arrangements, comparison can often be made without knowing the partial-QS values. For instance, for six-coordinate MAB$_5$, *cis*-MA$_2$B$_4$ and *trans*-MA$_2$B$_4$, the EFGs at atom M fall in the simple ratio $1: -1:2$ (see Fig. 4.6).

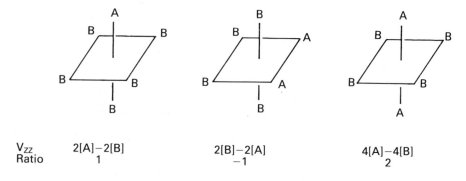

| V_{ZZ} | $2[A]-2[B]$ | $2[B]-2[A]$ | $4[A]-4[B]$ |
| Ratio | 1 | -1 | 2 |

Fig. 4.6 — Relative QS values for M in MAB$_5$, *cis*-MA$_2$B$_4$ and *trans*-MA$_2$B$_4$.

For systems with relatively low symmetry, the major axis of the EFG may not coincide with the symmetry axis. In such cases, it is usually sufficient to calculate values with each of the bond directions in turn as the *z*-axis, and to choose the largest. The commonest example is the *cis*-disubstituted octahedron considered above, MA$_4$B$_2$, where the EFG axis is the B–M–B direction and not the C_2 axis (the reason for this is that this choice of axes make the *x*- and *y*-axes equivalent, so that the asymmetry parameter is zero). The actual QS may be slightly larger than that calculated, owing to the effect of the asymmetry parameter; however, the maximum error introduced by ignoring this effect is 15%. It is, of course, possible to calculate an accurate value, but this involves estimating the values for all nine components of the EFG-tensor and diagonalizing the resulting matrix (see Problem 4.9).

4.4 INTERPRETATION

The simplest form of interpretation is by 'fingerprinting', i.e. to compare the experimental spectrum directly with those of known materials. If actual spectra for relevant compounds are not available, the parameters may be. Reference texts and reviews may be consulted (see Bibliography) and recent new data are collected and reported monthly in the *Mössbauer Effect Research and Data Journal*; the latter also operates as a computerized data base which can be searched on request. Alternatively, use may be made of the known trends in parameters for the isotopes, which are summarized below and in Table 4.3. Examples of the methods of interpretation are given in the problems at the end of the chapter.

4.4.1 Iron-57

Work with ^{57}Fe accounts for about half of all the data recorded. This is partly because the spectra are quite easy to obtain, and partly because iron occurs in a wide variety of chemical, biological and geological systems. Interpretation is complicated by the fact that $\Delta R/R$ is *negative*, so that increasing electron density at the nucleus results in a *decrease* in IS. Thus, increasing oxidation state and increasing donation by the ligands both give a decrease in IS. Another complication is that oxidation states of +2 to +4 may occur in high-spin or low-spin forms. Fortunately, the parameters for the various possibilities are fairly distinct (see Table 4.3), especially since the type of ligand involved (and hence the extent of its donation to iron) is usually known. The quadrupole splitting reflects the number and arrangement of d-electrons present: high-spin d^5 [iron(III)] and low-spin d^6 [iron(II)] have spherical and cubic distributions, respectively, so that there is no non-bonding electronic contribution to the EFG and quadrupole splitting arises entirely from the ligands. In the other cases, the contributions of the non-bonding and bonding electrons often partially cancel, and wide variations in QS are seen.

d^6 High-spin iron(II) is easily recognized from its high IS and (generally) large QS. High-spin iron(III) has lower IS and a small QS. For both systems, there is a systematic decrease in IS with decrease in co-ordination number, e.g. 0.2–0.3 mm s^{-1} between octahedral and tetrahedral co-ordination involving similar ligands. In all these cases, the IS decreases as the ligands become better donors to iron. When the iron–ligand bonds become sufficiently covalent there is a change to the low-spin state, and a further decrease in IS. The d-electrons are now delocalized by the bonding interactions, which decreases their shielding of the nucleus. For the same reason, the sensitivity of the IS to oxidation state is considerably decreased; for the same set of ligands, a low-spin iron(III) complex may have an IS only about 0.1 mm s^{-1} lower than the corresponding iron(II) complex (see Fig. 4.7).

The QS should be zero for high-spin iron(III) (d^5) and low-spin iron(II) (t_{2g}^6) in strict cubic symmetry (O_h or T_d). Lowering of the symmetry, either by distortion of the bond lengths or angles or by the presence of two or more different ligands, results

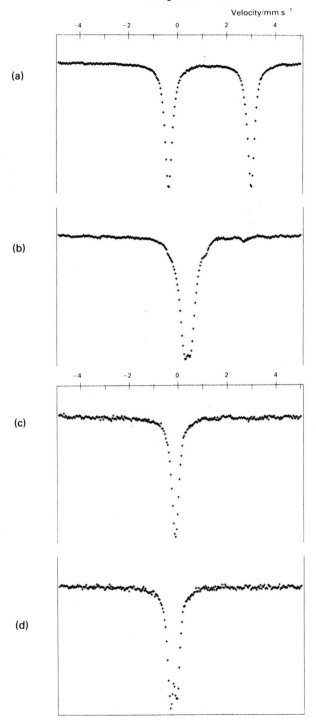

Fig. 4.7 — ^{57}Fe spectra for (a) FeSO$_4$.7H$_2$O, (b) Fe$_2$(SO$_4$)$_3$ (technical grade),
(c) K$_4$[Fe(CN)$_6$].3H$_2$O and (d) K$_3$[Fe(CN)$_6$].

in values within the ranges shown in Table 4.3. For high-spin iron(II), a large QS is expected from the unequal population of the t_{2g} orbitals when the symmetry is high and the bonding-electron contribution is small. Even with six identical ligands, the symmetry is rarely exactly cubic, there is nearly always a small loss in degeneracy of the t_{2g} orbitals. The magnitude of the QS then depends on which orbital contains the additional electron: an electron in the d_{xy}-orbital makes (minus) twice the contribution of an electron in the $d_{xz,yz}$ orbitals, and the QS may be up to about 4 or about 2 mm s^{-1} in these two cases. These values are usually upper limits, since the contributions of the bonding and non-bonding electrons are opposite in sign. If the t_{2g}-splitting is comparable with kT, the QS may show a temperature dependence, becoming smaller with increasing temperature (see the data in Table 4.8). The temperature-dependence of the IS shown by these complexes is also common: it is due to the second-order Doppler shift (i.e. the dependence of the gamma-energy on the square of the source-velocity).

For low-spin iron(II) complexes, scales of partial-IS and partial-QS values have been constructed (Tables 4.6, 4.7), so that estimates of the parameters can be made for any chosen set of ligands (see section 4.3.3). In principle, similar scales could be constructed for high-spin iron(III), but insufficient data are available.

High-spin iron(III) compounds in which the iron atoms are well separated sometimes show an asymmetric broadening of the spectrum owing to magnetic relaxation effects (see Fig. 4.8). Sharper spectra may, in principle, be obtained by raising the temperature, or the magnetic spectrum may be resolved by lowering the temperature (see section 4.5.3). The latter effect is also seen in magnetically ordered materials. Data for the analysis of fully magnetically split, six-line spectra are given in Table 4.5. The appearance of such spectra depend on the crystallites being large enough to maintain the spin-coupling between groups of iron atoms (i.e. larger than the normal magnetic domain size). For very small particles, anomalous spectra may be obtained, a phenomenon known as *super paramagnetism*.

With certain combinations of ligands, the ligand-field splitting energy may be comparable with the electron-pairing energy, so that two spin-states are in thermal equilibrium. This can readily be detected by measuring the Mössbauer spectrum at various temperatures (see Fig. 4.9 and Problem 4.3).

Table 4.8 — ^{57}Fe Mössbauer parameters for iron(II) complexes of pyridine (py)

Complex	T (K)	IS mm s^{-1}	QS mm s^{-1}	Complex	T (K)	IS mm s^{-1}	QS mms^{-1}
Fe(py)$_4$Cl$_2$	297	1.06	3.08	Fe(py)$_2$Cl$_2$	297	1.08	0.57
	78	1.18	3.49		78	1.21	1.25^a
Fe(py)$_4$Br$_2$	297	1.03	2.20	Fe(py)$_2$Br$_2$	297	1.03	0.82
	78	1.09	3.48		78	1.15	1.18
Fe(py)$_4$I$_2$	297	0.99	0.33	Fe(py)$_2$I$_2$	297	0.76	0.94
	78	1.11	0.53		78	0.86	1.33
Fe(py)$_4$(NCS)$_2$	297	1.05	1.54	Fe(py)$_2$(NCS)$_2$	297	1.02	2.60
	78	1.17	2.01		78	1.12	3.02

^{a} A phase-change occurs in this compound on cooling.
Data from B. F. Little and G. J. Long, *Inorg. Chem.*, **17** (1978) 3401.

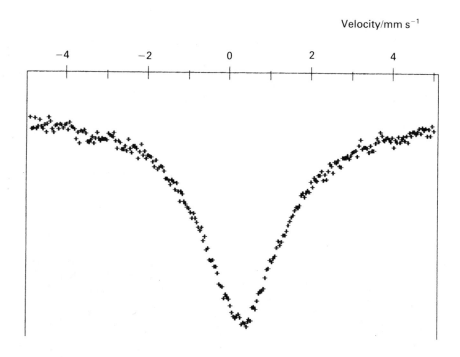

Fig. 4.8 — ^{57}Fe Mössbauer spectrum of (NH$_4$)Fe^{III}(SO$_4$)$_2$.12H$_2$O showing magnetic relaxation.

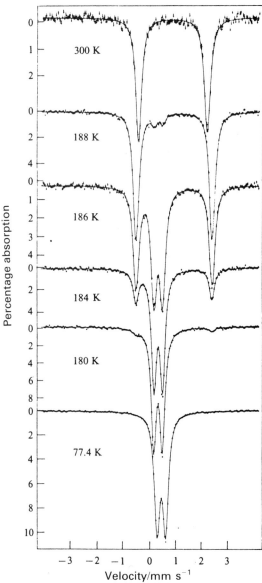

Fig. 4.9 — ^{57}Fe Mössbauer spectra for [Fe(NCS)$_2$(phen)$_2$] showing temperature variation of the spin-state. (Reproduced with permission from P. Gütlich, C. P. Kohler, H. Köppen, E. Meissner, E. W. Müller, and H. Spiering, *Trends in Mössbauer Spectroscopy*, P. Gütlich and M. Kalvius (Eds). University of Mainz, 1983.

Derivatives of iron pentacarbonyl have relatively low symmetry and a d^8 configuration, and substantial quadrupole splitting is always seen (1–3 mm s^{-1}, see Table 4.9). The same applies to organometallic derivatives, especially those of the Fe(π-C$_5$H$_5$)(CO)$_2$X-type; partial-IS scales have been proposed for such compounds (R. H. Herber, *Prog. Inorg. Chem.*, **8** (1967) 1).

Table 4.9 — ^{57}Fe Mössbauer data for some iron-carbonyl compounds (at 78 K)

$Fe(CO)_5$	-0.09	2.57	$Na_2Fe(CO)_4$	-0.18	0.00
$Fe_2(CO)_9$	$+0.17$	0.42	$NaFe(CO)_3(NO)$	-0.10	0.38
$Fe_3(CO)_{12}$	$+0.06$	0.07	$Fe(CO)_2(NO)_2$	$+0.06$	0.33
	$+0.12$	1.11	$Fe(CO)_4H_2$	-0.18	0.55
$Fe(CO)_4(PPh_3)$	-0.07	2.45	$Fe(CO)_4Cl_2$	$+0.05$	0.26
$Fe_2(CO)_6(\mu\text{-}PMe_2)_2$	-0.01	0.65	$Fe(CO)_4I_2$	$+0.06$	0.32
$Fe_3(CO)_9(PMe_2Ph)_3$	$+0.02$	0.57	$[CpFe(CO)_2]_2$	$+0.22$	1.89
	$+0.09$	1.15	$CpFe(CO)_2Cl$	$+0.24$	1.89

Data from compilation by R. V. Parish in *The Organic Chemistry of Iron* (see Bibliography).

4.4.2 Tin-119

There is a clear distinction in IS between the two major oxidation states of tin (Tables 4.3, 4.10). In addition, most tin(II) systems show a modest QS, owing to the stereochemically active lone pair. Inorganic tin(IV) compounds (i.e. those which do not contain a tin–carbon bond) show QS values up to about 1 mm s^{-1} when the ligands to tin are not all equivalent. The IS decreases with increase in co-ordination number, by about 0.5 mm s^{-1} between four- and six-coordination.

Table 4.10 — ^{119}Sn Mössbauer data (mm s^{-1})

	IS	QS		IS	QS
SnO^a	2.71	1.45	$SnO_2{}^b$	0.0	0.53
$SnF_2{}^a$	3.60	2.20	$SnF_4{}^b$	-0.47	1.66
$SnCl_2{}^a$	4.15	0.30	$SnCl_4{}^c$	0.80	0.00
$SnBr_2{}^a$	4.06	0.20	$SnBr_4{}^c$	1.10	0.00
$NH_4 SnF_3{}^a$	3.21	1.88	$(Et_4N)_2 SnF_6{}^d$	-0.36	0.00
$Et_4N SnCl_3{}^a$	3.46	1.13	$(Et_4N)_2 SnCl_6{}^d$	0.51	0.00
			$SnCl_4.bipy^e$	0.47	$ca\ 0$
			$SnCl_4(NCMe)_2{}^e$	0.38	0.91

Structures are:
^{a} pyramidal, ^{b}distorted octahedral, ^{c}tetrahedral, ^{d}octahedral, ^{e}cis-octahedral.

Organotin(IV) compounds containing up to three tin–carbon bonds all show substantial quadrupole splitting, the size of which depends mainly on the number and disposition of the organic groups but only slightly on their identity or that of the other ligands (Table 4.11). Inorganic ligands make only minor contributions to the quadrupole splitting, as is indicated by the partial-QS values of Table 4.7. These values may be used in conjunction with eqn (4.2) to calculate QS values for various structures and combinations of ligands; note that different scales are required for different co-ordination numbers (because of the changes in hybridization).

Table 4.11 — Quadrupole splitting ranges (mm s^{-1}) for organotin(IV) compounds

	RSnX$_3$	R$_2$SnX$_2$	R$_3$SnX
Tetrahedral	1.3–2.1	2.1–2.4	1.5–2.8
Trigonal-bipyramidal	1.6–2.4	2.9–3.7	2.6–3.9
Octahedral	1.6–2.4	1.7–2.2 *cis*	
		3.5–4.2 *trans*	

R = alkyl or aryl group, X = electronegative ligand.

Diorganotin(IV) compounds are usually either five- or six-coordinate. Ready distinction can be made between *cis* and *trans* six-coordination, provided the structures are reasonably regular, both from the quadrupole splitting (Table 4.11) and from the isomer shift (see below). However, the *trans* structures are often rather distorted, with C–Sn–C angles far from 180°. It is then difficult to distinguish between six- and five-coordination, both of which may give QS values of about 3.5 mm s^{-1}. Since the QS is governed mainly by the organic groups, it changes markedly with the bond angle, and good estimates of the QS for di-organotin(IV) compounds can be made from the formula

$$QS = -4[R][1 - (3/4) \sin^2 \alpha]^{1/2}$$

where α is the R–Sn–R bond angle, and [R] is the partial-QS value for the organic group. If the inorganic ligands have very high electronegativity, and, hence, positive partial-QS values (such as RSO$_3^-$), QS up to 4.5 mm s^{-1} may be found for $\alpha = 180°$.

Triorganotin(IV) compounds may be monomeric and four-coordinate or, in many cases, polymeric with five-coordination for tin; the latter gives the higher QS (Table 4.11).

It has not been found possible to construct satisfactory scales of partial-IS values, except for short series of strongly related compounds. However, *trans* C–Sn^{IV}–C systems give significantly higher IS than their *cis* counterparts (1.2–1.6 versus 0.8–1.0 mm s^{-1}). Mono- and tri-organotin(IV) compounds give IS-values of 0.7–1.0 and 1.2–1.5 mm s^{-1}, respectively.

4.4.3 Antimony-121
Owing to the high nuclear spins involved ($I_g = 5/2$, $I_{ex} = 7/2$), the spectra are complex and the many lines (8 or 12) composing a quadrupole-split spectrum are always heavily overlapped (see Fig. 4.3(c)). Computer fitting of the data is essential and, for values of the QCC less than about 15 mm s^{-1} (450 MHz), the transmission integral should be used.

When the difference in sign of $\Delta R/R$ and the higher sensitivity of the IS to chemical factors are taken into account, the interpretation of data for antimony compounds is very similar to that for tin (*q.v.*). Thus, the two major oxidation states are well separated in IS (Table 4.3), nearly all antimony(III) compounds show substantial quadrupole splitting, and for antimony(V) the partial-QS treatment can be successfully used (Table 4.7). For organo-antimony(V) compounds, the parameters are determined primarily by the organic groups.

4.4.4 Iodine-127 and Iodine-129

Both available isotopes have 5/2, 7/2 spin states, and the Mössbauer spectra are complex. The naturally occurring isotope, ^{127}I, gives spectra which are much more poorly resolved than those of the very expensive, mildly β-active ^{129}I. The use of the latter is essential if accurate values of the QCC are required when it is expected to be less than about 700 MHz. Note also that the two isotopes differ in the sign of $\Delta R/R$, so that the IS scales work in opposite directions. In order to facilitate comparison with NQR data, it is usual to quote the QCC in MHz and to convert values for ^{129}I to the ^{127}I-scale, using the relationships

$$e^2qQ^{127}/\text{MHz} = 32.58\ e^2qQ^{129}/(\text{mm s}^{-1})$$
$$= 46.46\ e^2qQ^{127}/(\text{mm s}^{-1})$$

The five oxidation states of iodine give distinct sets of parameters (Table 4.3, Fig. 4.10). For iodide ions ($5s^25p^6$) and iodine(VII) (formally $5s^05p^0$) the QCC is zero or very small, but the IS ranges are very different. Iodine(I) compounds of type IX give QCC values which become increasingly negative as X becomes more electronegative (eQ is negative for both isotopes). Both iodine(III) and iodine(V) normally give low-symmetry structures, with stereochemically active lone pairs. Substantial positive QCC values are found, due predominantly to the effects of the lone pairs: QCC values for iodine(III) are therefore approximately twice those for iodine(V). In both cases, increasingly covalent bonding leads to diminution of the QCC value, so that purely inorganic compounds give high values and organic derivatives low values. Representative data are given in Table 4.12.

4.4.5 Iridium-193

Since $\Delta R/R$ is positive, the IS is expected to increase with increasing oxidation state and with increasing donation by the ligands. Iridium(III) (t_{2g}^6) is normally six-coordinate, and shows substantial QS only when more than one type of ligand is present. Partial-IS and partial-QS values are available (Tables 4.6, 4.7). The IS range for iridium(III) quoted in Table 4.3 is much wider than that for the other oxidation states simply because a wider range of compounds has been examined.

Iridium(I) is usually four-coordinate and square-planar [*cf.* gold(III), in the next section]. The d^8 configuration gives a negative contribution to the QS which is outweighed by that from the ligands; the QS therefore increases as the iridium–ligand bonds become increasingly covalent. From the data available at present, it appears that systematic changes of IS are smaller than the experimental errors.

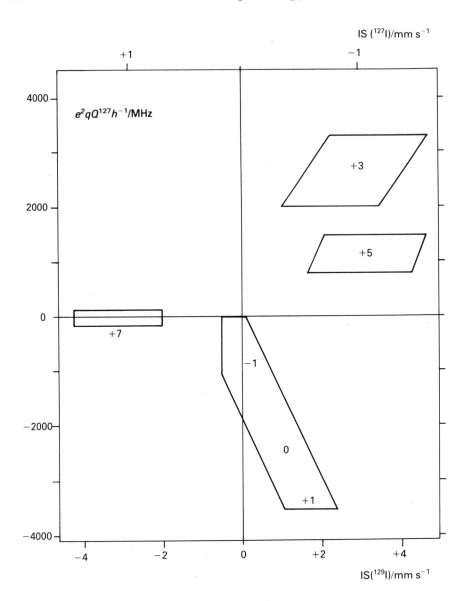

Fig. 4.10 — IS–QCC correlation diagram for the iodine isotopes, showing characteristic regions for the different oxidation states.

4.4.6 Gold-197

Gold(III) is isoelectronic with iridium(I), and also forms predominantly square-planar complexes. Both Mössbauer parameters are sensitive to the chemical environment of the gold atom, increasing with increasing covalency of the bonds (Fig. 4.11). For the least covalent ligands (e.g. Cl^-), the ligand-contribution to the

Table 4.12 — ^{127}I and ^{129}I Mössbauer data

	^{127}I		^{129}I	
	IS(ZnTe) mm s^{-1}	QCC MHz	IS(ZnTe) mm s^{-1}	QCC[a] MHz
I_2	-0.58	-2238	0.83	-2154
ICl	-0.62	-2868	1.73	-3131
IBr			1.23	-2892
CI_4	-0.35	-2160	0.20	-2102
CHI_3	-0.21	-2060	0.52	-2065
CH_2I_2	-0.14	-1920		
K ICl$_4$	-1.30	3097		
I_2Cl_6	-1.27	3028	3.50	3060
$I(OAc)_3$			3.60	2394
IF$_5$	-1.34	1170		
PhIO$_2$	-0.88	1244		

[a] On the ^{127}I-scale (see text).
From the compilation by R. V. Parish, *Mössbauer Spectroscopy Applied to Inorganic Chemistry*, Vol. 2 (see Bibliography).

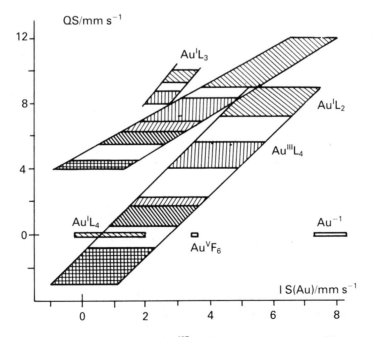

Fig. 4.11 — *IS–QS* correlation diagram for ^{197}Au. The hatchings represent different types of ligands: \\\, P and C-donors; |||, S-donors; ///, N-donors; \\\, anionic halide complexes; cross-hatching, neutral halides.

QS is very similar, and opposite in sign to, that from the d^8-electron configuration, and the nett QS value is close to zero. Mixed-ligand complexes, AuX_2Y_2, give parameters which are roughly the average for AuX_4 and AuY_4, but insufficient data are available to construct partial-IS or partial-QS scales. Geometric isomers cannot be distinguished: since the EFG is perpendicular to the co-ordination plane, both *cis* and *trans* isomers will give the same QS. The rather rare cases of higher co-ordination number can be recognized by diminution of both parameters. Representative data are given in Table 4.13.

Table 4.13 — ^{197}Au Mössbauer data (mm s^{-1})

	IS(Au)	QS		IS(Au)	QS
$AuCl_2^-$	1.84	6.28	$AuCl_4^-$	2.06	1.27
$AuBr_2^-$	1.66	6.35	$AuBr_4^-$	2.02	1.34
$Au(SMe_2)_2^+$	3.43	7.56	$Au(SCN)_4^-$	2.84	2.04
$Au(py)_2^+$	4.04	7.88	$Au(mnt)_2^{-a}$	4.13	2.33
$Au(PPh_3)_2^+$	5.24	9.53	$AuMe_2(S_2CNBu_2)$	5.20	5.11
$Au(CN)_2^-$	3.99	10.01	$Au(CN)_4^-$	5.4	6.9
$Au(C_6H_4NMe_2)_2^-$	6.86	12.01	$AuMe_3(PPh_3)$	5.91	8.87
$AuCl(SMe_2)$	4.44	8.20	$AuMe_2Cl(PPh_3)$	5.28	6.32
$AuCl(PPh_3)$	4.08	7.43			
$Au(PPh_3)_3^+$	2.99	9.47			
$AuCl(PPh_3)_2$	2.35	8.22			
$Au(PPh_3)_4^+$	− 0.17	0.00			
$Au(PPhMe_2)_4^+$	1.98	0.00			

Values for ions are averages over the known salts.
^{a} mnt, dianion of maleonitrile.
From the compilations by R. V. Parish *Mössbauer Spectroscopy Applied to Inorganic Chemistry* (see Bibliography).

Gold(I) (d^{10}) mostly gives linear two-coordination, so that a QS is expected even when the ligands are identical ($D_{\infty h}$). Increased covalency gives a systematic increase in both Mössbauer parameters (Fig. 4.11), and it is usually possible to identify the ligands from the observed values. Scales of partial-IS and partial-QS values are available (Tables 4.6, 4.7). Increase in co-ordination number gives a decrease in IS by 1–2 mm s^{-1} (three-coordination) or 2–4 mm s^{-1} (four-coordination). For the same ligands, the QS for AuL_3 is expected to be very similar to that for AuL_2, while that for AuL_4 (T_d symmetry) should be zero. Complexes of the type Au^IX_3Y give small QS values, but are easily recognized by their low IS. Representative data are given in Table 4.13.

4.5 COMPLICATIONS

4.5.1 Unusual intensities

For the 1/2, 3/2 isotopes, quadrupole-split doublets are expected to be symmetrical, i.e. the two lines should have equal width and equal intensity. However, it is not uncommon for the two peaks to be rather different, especially for [57]Fe or [119]Sn spectra. The most probable cause for this is **texture** in the sample, that is, the crystallites of the sample have become aligned during packing of the cell; this is very likely when the crystals are plates or needles. A similar effect may occur when a liquid sample is frozen in a cell. The underlying reason is that the intensities of the transitions to the $m_I = \pm 3/2$ and $\pm 1/2$ levels have different dependence on the angle between the principal EFG axis and the γ-beam. In a randomly oriented powder, all angles are present, and the average intensities are the same. In a completely oriented sample, the intensity ratio can be as low as $3:5$ ($\pm 1/2 \leftrightarrow \pm 1/2 : \pm 1/2 \leftrightarrow \pm 3/2$) when the angle is 90°, and up to $6:2$ when the angle is 0°. If the crystal-structure is known, this provides a simple method of determining the sign of the EFG.

For a similar reason, the intensities of six-line magnetically-split spectra are not always in the theoretical $3:2:1:1:2:3$ ratio expected for randomly oriented crystallites. If the sample is oriented, or if an external magnetic field is being applied, the ratio changes from $3:4:1:1:4:3$ when the angle between the magnetic field and the γ-beam is 90° to $3:0:1:1:0:3$ when the two are parallel.

Another possible reason for lack of symmetry in a quadrupole-split doublet is the **Goldanskii–Karyagin** effect which can, in principle, occur whenever the Mössbauer atom is at a low-symmetry site. It arises from anisotropy in the recoil-free fraction, that is, when the Mössbauer atom has much more vibrational amplitude in one direction than another. This effect is seldom convincingly seen but can be distinguished from a texture effect by running the spectrum at a different temperature: texture effects will remain whereas the Goldanskii–Karyagin effect becomes less pronounced as the temperature is lowered.

For the more complex spin-states, the theoretical intensity ratios for quadrupole-split spectra are as given in Table 4.4 only for 'thin' samples. As the thickness of the sample increases, Beer's law is no longer obeyed, the strong lines become less intense than expected, and the weaker lines become more intense. The line shape may also deviate significantly from the theoretical Lorentzian. Since most fitting programs apply a weighting to the data which is the reciprocal of the number of counts in each channel, the strongest lines have the greatest influence on the derived parameters. If the intensities are anomalous, inaccurate values will be obtained. Under these circumstances, it is necessary to use a fitting program which employs a transmission integral treatment allowing for the effects of the change in line shape with increasing thickness. The sample thickness now becomes one of the variable parameters of the fitting procedure. Such programs are expensive of computer time, and should be used only for the final refinement of data.

4.5.2 Non-zero asymmetry parameter

The asymmetry parameter, η, appears in expressions for the energies of the various m_I states in the presence of an EFG. For $I = 3/2$ [$m_I = \pm 3/2, \pm 1/2$], there is a simple dependence and the QS is given exactly by $V_{zz}(1 + \eta^2/3)^{1/2}$. For more complex spin states, no simple analytical expressions are possible. However, the energies of the m_I states can be expressed as a power series in η^2 which converges rapidly (Table 3.2). When this procedure is applied to both Mössbauer spin states, expressions for the transition energies can be derived, examples of which are shown in Table 4.14.

However, since the relative intensities of the lines also change, it is better to use a computer program specially designed to calculate the energies and intensities correctly, by solving the Hamiltonians for the spin states. Such procedures can be embodied in fitting programs.

Table 4.14 — Line positions for non-zero asymmetry parameter

Line	^{127}I			^{129}I		
	A	B	C	A	B	C
1	+ 0.1038	+ 0.1883	− 0.2820	+ 0.4149	− 0.0936	+ 0.2624
2	+ 0.0820	− 0.0483	+ 0.0516	+ 0.2720	− 0.0126	− 0.0032
3	+ 0.0397	− 0.0447	+ 0.1838	+ 0.1170	+ 0.2122	− 0.3162
4	− 0.0255	− 0.0132	− 0.0010	+ 0.0578	+ 0.0088	+ 0.0012
5	− 0.0462	+ 0.0244	− 0.1669	+ 0.0456	− 0.0184	+ 0.2026
6	− 0.1104	− 0.2086	+ 0.2789	− 0.0676	+ 0.0394	− 0.1991
7	− 0.2180	+ 0.0128	+ 0.0030	− 0.0972	+ 0.0626	− 0.0632
8	− 0.3462	+ 0.0855	− 0.2356	− 0.1390	− 0.2201	+ 0.3196

Line position/mm s^{-1} = e^2qQ [A + Bη^2 + Cη^4].

4.5.3 Magnetic ordering and magnetic relaxation

Unpaired electrons should give a magnetic field at the nucleus. However, at normal temperatures the effects of thermal vibration usually average this field to zero, and no magnetic splitting is seen in the Mössbauer spectrum. If the paramagnetic atoms are not well separated in the crystal lattice (the sample is magnetically concentrated), the electron spins on adjacent atoms may become coupled together, resulting in magnetic ordering. This effect is common in iron(III)-oxide systems, and quite common for co-ordination compounds at low temperatures. The sample is now ferromagnetic (or sometimes antiferromagnetic, if the spins on adjacent atoms are coupled antiparallel). Within a given region, the electronic spins are all aligned in a particular direction; their magnetic field at the nucleus is no longer zero, and may amount to many tens of Tesla. The Mössbauer spectrum will now show magnetic splitting. If the temperature is raised, thermal vibrations will begin to oppose the ordering, and there will come a point at which the ordering breaks down and the sample becomes paramagnetic again. This is the Néel temperature, which is often sharply defined, and its value may be used to characterize the magnetic phase. The spectrum now appears as an asymmetric doublet, where the two lines have different widths as well as different intensities.

Just below the Néel temperature the spectrum shows features of both the magnetically ordered and the paramagnetic systems, with each spectrum broadening and merging into the other. Often the baseline appears to be highly curved. This is the phenomenon of **magnetic relaxation** and it is sometimes seen in magnetically dilute systems. High-spin iron(III) is particularly prone to show magnetic relaxation when the iron atoms are highly dilute, e.g. in a frozen solution or when ion-exchanged into a host lattice (see Fig. 4.8). Two remedies are possible. The relaxation effects may be removed by raising or lowering the temperature substantially, putting the sample firmly into either the paramagnetic or the ordered region, or a special computer program may be used which takes the relaxation temperature and the magnetic field at the nucleus as two of its parameters.

4.5.4 Recoil-free fractions

In principle, every different type of site for a Mössbauer atom will have a unique value for the f-factor, which depends on the mass of the recoiling unit and the tightness of its binding in the lattice. In practice, such differences are often quite small, especially in metallic or ionic lattices. If this is the case, the relative areas of the subspectra can be taken as a reliable guide to the relative concentrations of the various sites. Occasionally this procedure can give rise to serious error, particularly for samples consisting of molecular species or complex ions. The recoiling unit then often approximates to the whole molecule or ion and, if these have appreciably different masses, their f-factors can be significantly different. This is especially true when there are differences in co-ordination number or oxidation state for the Mössbauer atom. For instance, the spectrum shown in Fig. 4.12 consists of two quadrupole doublets of equal area. X-ray analysis revealed that there were two gold(I) atoms present for every gold(III) atom. This is an extreme example, but serves to illustrate the possible dangers of interpretation of spectrum area.

Recoil-free fractions can be determined by careful measurement of the same sample over a range of temperatures, or by mixing the sample with a known amount of a substance whose f-factor is known.

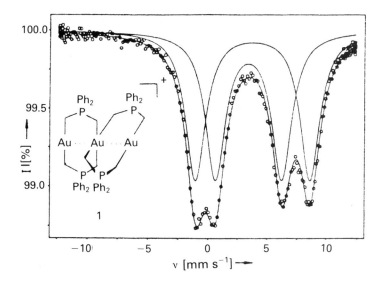

Fig. 4.12 — ^{197}Au Mössbauer spectrum for $[\{Ph_2P(CH_2)_2\}_4Au^I_2Au^{III}]I$ (Reproduced with permission from H. Schmidbaur, C. Hartmann, and F. E. Wagner, *Angew. Chem. Engl. Ed.*, **16** (1987) 1148.)

BIBLIOGRAPHY

G. J. Long (Ed.), *Mössbauer Spectroscopy Applied to Inorganic Chemistry*. Plenum Press, New York, Vol. 1, 1984; Vol. 2, 1987; Vol. 3, 1989. Review chapters on a wide variety of chemical topics.

D. P. E. Dickson and F. J. Berry (Eds), *Mössbauer Spectroscopy*. Cambridge University Press, 1986. Good for bonding, ^{57}Fe, and magnetism.

P. Gütlich, R. Link, and A. Trautwein, *Mössbauer Spectroscopy and Transition Metal Chemistry*. Springer-Verlag, Berlin, 1978. Useful reviews of data for transition metal compounds.

T. C. Gibb, *Principles of Mössbauer Spectroscopy*. Chapman & Hall, London, 1977. Concise treatment of theory, with examples.

M. G. Bancroft, *Mössbauer Spectroscopy in Inorganic Chemistry*, McGraw-Hill, New York, 1973. A good, readable treatment of theory for chemists, with many examples, especially for iron-containing minerals.

N. N. Greenwood and T. C. Gibb, *Mössbauer Spectroscopy*. Chapman & Hall, London, 1971. Excellent, comprehensive treatment. Data and examples now rather dated.

G. K. Wertheim, *Mössbauer Effect — Principles and Applications*. Academic Press, New York, 1964. Very readable account of basic theory.

G. K. Shenoy and F. E. Wagner (Eds), *Mössbauer Isomer Shifts*. North-Holland, Amsterdam, 1978. Much data and theory for wide variety of isotopes.

J. Stevens and G. K. Shenoy (Eds), *Mössbauer Spectroscopy and its Chemical Applications*. American Chemical Society, Washington, DC, 1981. Special techniques, minerals, catalysts (all mainly iron).

G. M. Bancroft and R. H. Platt, *Adv. Inorg. Chem. Radiochem.*, **15** (1972) 59. 'Mössbauer Spectra of Inorganic Compounds: Bonding and Structure'. Data and interpretations for iron, tin and iodine compounds, to 1971.

Data compilation

Mössbauer Effect Reference and Data Journal, published ten times per year by the Mössbauer Effect Data Center, University of North Carolina, Asheville, North Carolina 28804–3299, USA. Current-awareness compilation of data and references for most recent publications. The data bank can be computer-searched.

PROBLEMS

Answers on pp. 163–167

P4.1 The mineral howieite, approximately $NaMn_3Fe_9Si_{12}O_{31}(OH)_{13}$, gives a Mössbauer spectrum which shows the presence of iron in two sites (Fig. 4.13). Analyse the spectrum, deduce the oxidation states present and use the areas of the sub-spectra to derive their approximate relative proportions.

[Note: area = (peak height) × (width at half-height)]

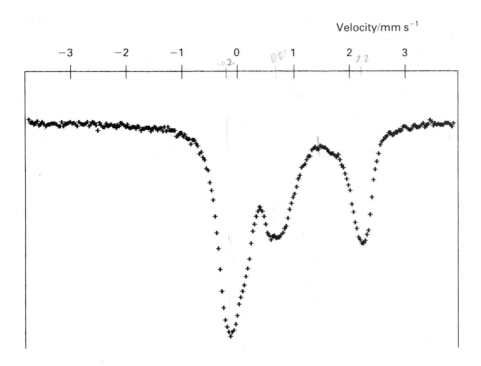

Fig. 4.13 — ^{57}Fe Mössbauer spectrum of the mineral howieite.

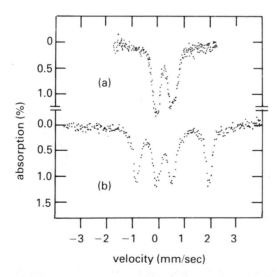

Fig. 4.14 — ^{57}Fe Mössbauer spectra for (a) the oxidized and (b) the reduced form of *Scenedesmus* ferridoxin. (Reproduced with permission from C. E. Johnson, *J. Appl. Phys.*, **42** (1971) 1325.)

P4.2   The ^{57}Fe Mössbauer spectra of the two-iron ferredoxin (iron–sulphur protein) *Scenedesmus* is shown for the oxidized and reduced states of the protein (Fig. 4.14). How many electrons are involved in the redox process?

P4.3   The spin-state of the complex *cis*-[Fe(NCS)$_2$(phen)$_2$] depends on the temperature. Which might be expected to be the low-temperature form? Do the spectra shown in Fig. 4.9 confirm your suggestion?

P4.4   The three spectra shown in Fig. 4.15 correspond to Fe(CO)$_5$, Fe$_2$(CO)$_9$, and Fe$_3$(CO)$_{12}$, which have structures (**4.1–III**). Assign each spectrum to the appropriate structure.

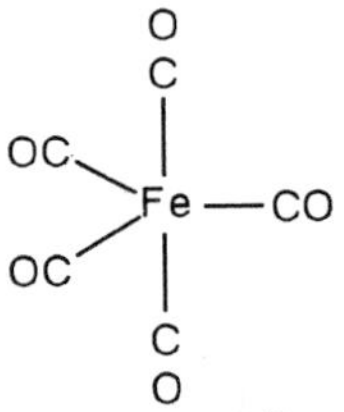

4.I 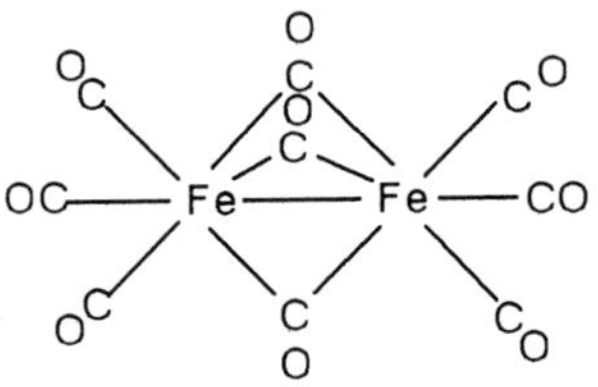   4.II 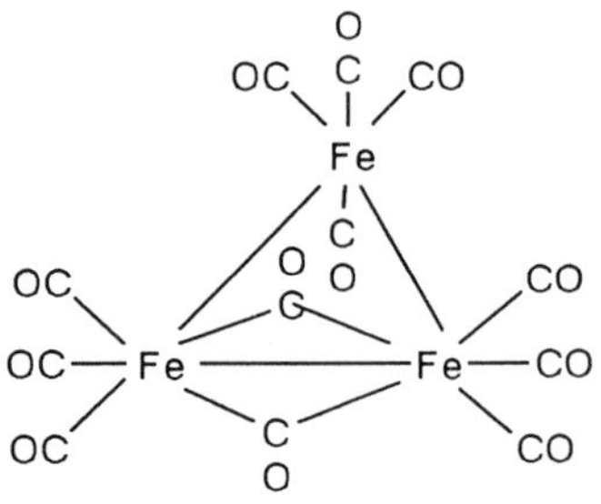   4.III

P4.5   Complexes of composition FeX$_2$(pyridine)$_2$ may be monomeric, with four-coordination for iron (**4.IV**), or polymeric with six-coordination (**4.V**). From the data given below for X = Cl and I, deduce which complex is polymeric.

Complex	IS/mm s^{-1}	QS/mm s^{-1}
FeCl$_2$ (pyridine)$_2$	1.21	1.25
FeI$_2$ (pyridine)$_2$	0.86	1.33

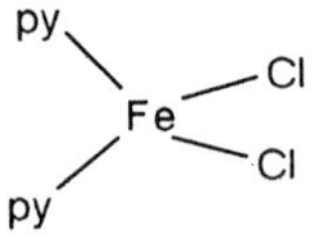

4.IV 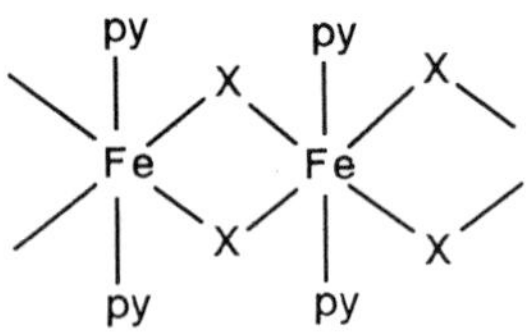   4.V

P4.6   From the spectrum of Fe$_2$O$_3$ shown in Fig. 4.5(b), estimate the strength of the

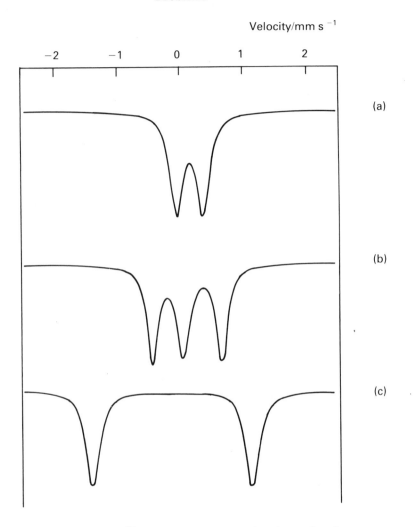

Velocity/mm s^{-1}

(a)

(b)

(c)

Fig. 4.15 — Simulated ^{57}Fe Mössbauer spectra for three iron carbonyls.

magnetic field at the iron nuclei. Do you think this spectrum shows any quadrupole spitting? [Use Table 4.5.]

P4.7 Tin(II) can be ion-exchanged into zeolites, but it is difficult to avoid aerial oxidation during the process. It is also difficult to determine the oxidation state of the tin chemically. Use the two ^{119}Sn Mössbauer spectra shown to estimate the relative amounts of tin(II) and tin(IV) in the two samples (Fig. 4.16). What assumptions do you make in this estimation?

P4.8 ^{119}Sn Mössbauer data for the tri-organotin(IV) formates, R$_3$SnO$_2$CH, are shown below. Deduce whether these compounds have monomeric or polymeric structures in the solid state (**4.VI** or **4.VII**).

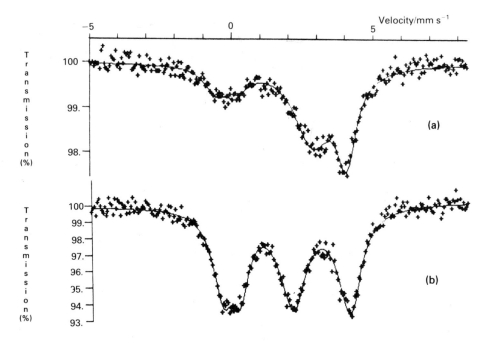

Fig. 4.16 — ^{119}Sn Mössbauer spectra of two tin(II) ion-exchanged zeolites.

R	IS/mm s^{-1}	QS/mm s^{-1}
Me	1.32	3.59
Ph	1.48	3.48

P4.9 NMR evidence suggests that in the compound SnMeBr(aph)$_2$ one bidentate ligand does not chelate (structure **4.VIII**, aph = C$_6$H$_4$CH$_2$NMe$_2$). Given that its QS is 2.90 mm s^{-1}, and that of SnMe$_2$Br(aph) (**4.IX**) is 2.90 mm s^{-1}, decide whether the Mössbauer and NMR data are self-consistent.

Derive a value for the partial-QS of the N-end of aph from the QS of **4.IX**. Assume that partial-QS of the C-end is the same as for C$_6$H$_5^-$.

Assuming that the structure may not be the same in solid and solution, what other structures should be considered? Can any be eliminated?

Given that the asymmetry parameter can be shown to be very small, which of all the possible structures is most likely to be correct?

4.VIII   4.IX   4.X

P4.10   The ^{121}Sb spectrum of Ph_3Sb shows substantial quadrupole splitting (QCC = + 16.2 mm s^{-1}). The platinum(II) complexes $[PtX_2(SbPh_3)_2]$ (X = Cl. I. NO_2) show much smaller values, e.g. + 9.3 mm s^{-1} for X = NO_2. To what may this decrease by attributed?

Cf. C. A. McAuliffe, I. E. Niven, and R. V. Parish, *J. Chem. Soc. Dalton Trans.* (1977) 1901.

P4.11   The ^{129}I parameters for I_2Cl_6 and $I_2Cl_4Br_2$ are shown below. Which of the possible structures for the latter is consistent with these data? [Note: iodine(III) is usually square-planar.]

	IS^{129}/mm s^{-1}	QCC/MHz
I_2Cl_6	+ 3.50	+ 3060
$I_2Cl_4Br_2$	+ 3.48	+ 3040
	+ 2.82	+ 2916

P4.12 ^{197}Au Mössbauer spectra for the compounds $[Au(PPh_3)_n]ClO_4 (n = 2,3,4)$ are shown (Fig. 4.17). Deduce which value of n applies to each spectrum.

P4.13   When an alcoholic solution of $Na[AuCl_4]$ is treated with triethylphosphine, a white solid product is obtained which shows the following ^{197}Au Mössbauer parameters: IS, 3.78 mm s^{-1}; QS, 7.08 mm s^{-1}. What is the oxidation state of the gold in this material? Can you suggest a formula?

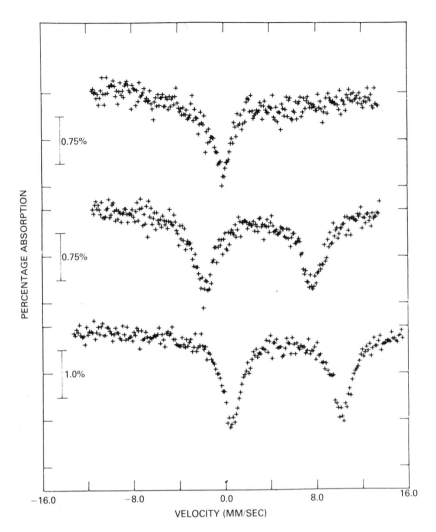

Fig. 4.17 — ^{197}Au Mössbauer spectra for $[Au(PPh_3)_n]ClO_4$.

ANSWERS

A4.1 The asymmetry of the spectrum clearly shows the presence of two components. Since doublets are expected to be symmetrical, two analyses are possible: (a) a
singlet at about -0.2 mm s^{-1} plus a doublet with IS ca 1.6, QS ca 1.7 mms^{-1}, or (b)
two doublets whose lower-velocity peaks overlap, IS ca 0.5 and 1.2 mm s^{-1}, QS 0.6,
2.6 mm s^{-1}. Reference to Table 4.3 shows that (a) is unreasonable, while (b)
corresponds to high-spin iron(II) and iron(III). The two resolved peaks are thus the
high-velocity components of the two doublets. Since these are approximately equal
in area, iron(II) and iron(III) are present in approximately equal amounts.

A4.2   The spectrum of the oxidized form shows only one doublet, with parameters consistent with tetrahedrally (S_4) coordinated iron(III). The reduced form shows two doublets, one of which is very similar to that of the oxidized form, and the other has a much higher IS and QS. The second doublet must therefore correspond to iron(II), so that only one electron is involved, and only one iron atom is reduced. [Careful inspection of the first spectrum reveals some broadening of the high-velocity component, suggesting that the two iron(II) atoms are not equivalent even in the oxidized form.]

A4.3   When iron(II) goes from high-spin to low-spin, an electron is transferred from the e_g orbital to the t_{2g} orbital. This will result in a decrease in the radius of the atom. As the temperature is reduced, the lattice will contract, favouring the adoption of the low-spin configuration. The parameters of the two doublets clearly confirm that this is happening (cf. Table 4.3).

A4.4   Only $Fe_3(CO)_{12}$ contains non-equivalent iron atoms. It therefore corresponds to spectrum (b). The wide doublet corresponds to two iron atoms, and the central, very close doublet, to the unique atom. Spectra (a) and (c) differ greatly in QS. The iron atom in $Fe(CO)_5$ occupies a site of low symmetry (D_{3h}) and would be expected to show a large QS; spectrum (c) must therefore be assigned to $Fe(CO)_5$. In confirmation, the site in $Fe_2(CO)_9$ is fairly similar to that of the two in $Fe_3(CO)_{12}$, but is less asymmetric; it shows a similar IS but a smaller QS.

A4.5   A change in coordination number affects the IS, principally through the change in hybridization. For tetrahedral, sp^3 hybridization there is less shielding of the s-electron density than for the octahedral case, d^2sp^3. For ^{57}Fe, an increase in s-electron density gives a decrease in IS; therefore the iodide has tetrahedral coordination.
B. F. Little and G. J. Long, *Inorg. Chem.*, **17** (1978) 3401.

A4.6   The line positions are approximately -8.1, -4.7, -1.2, $+1.4$, $+4.9$ and $+8.5$ mm s^{-1}. From Table 4.5, the outermost lines are separated by $0.33(B/T)$ mm s^{-1}. Hence,

$$0.33(B/T) \text{ mm s}^{-1} = 16.6 \text{ mm s}^{-1}$$

and

$$B = 50 \ T.$$

Quadrupole splitting would be reflected in a difference in the separations between the two left-most lines and the two right-most lines, which is clearly present.

A4.7   It is first necessary to assign the signals. Tin(IV) gives IS close to zero, and the QS is expected to be small in an oxide-type lattice. Tin(II) should give a higher IS and some QS (*cf.* Table 4.3). In spectrum (a), the tin(IV) signal is the low-intensity doublet with an IS close to zero; this signal accounts for 10–20% of the total area of the spectrum. In (b), the two sub-spectra are much closer in intensity, but the two peaks of the tin(IV) spectrum are more closely overlapped than the tin(II) peaks. The relative areas are therefore in a ratio rather less than $1:1$, but probably a little more than $1:2$, i.e. the tin(IV) area is 35–45% of the total.

To say that these area ratios represent the relative amounts of the two forms of tin involves the assumption that the recoil-free fraction is the same for both oxidation states. This is unlikely. [Probably, tin(IV) will have the greater f-factor, so the area ratio over-estimates the amount of tin(IV).]

A4.8 For organotin(IV) compounds structural diagnoses are best made from the QS values, which respond well to an additive treatment (section 4.2.3). The partial-QS value for a formate anion will be close to zero for either mode of bonding. The values for methyl and phenyl groups are about -1.2 mm s^{-1} (slightly greater for tetrahedral than trigonal-bipyramidal coordination). The tetrahedral R_3SnO system would therefore give a QS of $2[O] - 2[R] = ca\,(-)2.4$ mm s^{-1}. In the polymer, the R-groups lie in the equatorial plane of the trigonal bipyramid, at 90° to the z-axis, with the bridging formates lying on the axis. The QS would then be $4[O] - 3[R] = ca$ 3.6 mm s^{-1}. The experimental data clearly agree with the latter.
Cf. K. C. Molloy, K. Quill, and I. W. Nowell, *J. Chem. Soc., Dalton Trans.*, (1987) 101.

A4.9 Since **4.VIII** and **4.IX** differ only in the replacement of a methyl group by C_{aph}, the QS values are expected to be almost the same. They should differ by the difference between $[CH_3^-]$ and $[C_{aph}]$, $ca\,0.01$ mm s^{-1}, but may have different values of η. The Mössbauer data are thus not inconsistent with the structure proposed from the NMR data.

For structure **4.IX**, the QS is calculated as $2[N_{aph}] + 2[Br^-] - [C_6H_5^-]^{tbe}$ $- 2[CH_3^-]tbe = 2[N_{aph}] + 2.81$ mm s^{-1}. Since this must equal $+2.90$ mm s, $[N_{aph}] =$ $+0.04$ mm s^{-1}. Hence, $QS(\mathbf{4.VIII}) = 2[N_{aph}] + 2[Br^-] - 2[C_6H_5^-] - [CH_3^-] =$ 2.88 mm s^{-1}.

Six-coordinate structures in which the aph ligands are both chelated should be considered. Since $[Br^-]$ and $[N_{aph}]$ are very small, the QS will be governed by the disposition of the Sn–C bonds. A *fac* structure will always give a very small QS (it is zero for *fac*-SnA_3B_3, whatever the identity of A and B), cf. **4.X**. Such structures can therefore be eliminated. Four structures with *mer* geometry are possible (**4.XI–XIV**), for which the choice of axes is not obvious. This is a case in which it is necessary to perform the calculation for each axis in turn, and to select the maximum numerical value. The results are shown below, from which it is evident that all give about the same QS, and would also be consistent with the observed QS.

It is also evident that for each of the *mer* structures, η will be very large, since one of the components of the EFG is almost zero. Since, it is observed that η is actually quite small (0.14), **4.VIII** is the actual structure.
Cf. R. Barbieri, A. Silvestri, G. van Koten, and J. G. Noltes, *Inorg. Chim. Acta* **40**, (1980) 267.

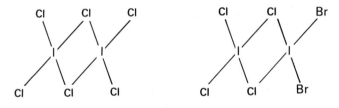

	4.XI	4.XII	4.XIII	4.XIV
V_{zz}	+ 3.09	2.97	+ 3.09	− 3.09
V_{yy}	− 2.85	− 2.85	− 3.09	+ 2.97
V_{xx}	− 0.24	− 0.12	0.00	+ 0.12

A4.10   The quadrupole splitting of Ph_3Sb is largely due to the lone pair of electrons (Q_{gd} is negative). When the ligand coordinates, the lone pair is donated to (shared with) the metal atom. The average distance of this pair of electrons from the antimony nucleus therefore increases and its contribution to the EFG decreases.

A4.11   Several structures are possible for $I_2Cl_4Br_2$, of which those in which the iodine atoms are equivalent are ruled out by the observation of two distinct sub-spectra. The only remaining structure is $Cl_2I(\eta - Cl)_2IBr_2$, which is confirmed by the fact that one sub-spectrum has parameters very similar to those of I_2Cl_6.

A4.12   The expected coordination for gold(I) is linear ($n = 2$), trigonal ($n = 3$) or tetrahedral ($n = 4$). The last of these should give no EFG at the gold nucleus, and therefore corresponds to the singlet spectrum. For the remaining two, the QS values are expected to be similar (as is found), but the IS should be substantially smaller for $n = 3$ than for $n = 2$.

A4.13   Using Fig. 4.11, the parameters lie in the gold(I) region. They appear to be mid-way between those expected for gold(I) coordinated to two halide ions or to two phosphine ligands. The product is therefore [$AuCl(PEt_3)$].

5

Electron paramagnetic resonance

Electron paramagnetic resonance (EPR) is different from all the other techniques discussed previously in that, as is obvious from the name, it is electrons which are of interest rather then nuclei. Equally obviously, the systems examined must be paramagnetic, i.e. it is unpaired electrons which are important. EPR thus applies to compounds of the various transition series and to free radicals. It should also be said that it is a technique from which it is a little more difficult to obtain detailed information than those considered previously, particularly about transition-metal systems. The extraction of parameters from the spectra is relatively easy, so that 'fingerprinting' can be done. To go further than that often requires some knowledge about the EPR behaviour of the particular ion. It should also be said that it is not always easy to obtain the spectra; some systems give very broad, weak signals and may require the use of liquid-helium temperatures.

Nonetheless, EPR is a useful complement to NMR, which cannot normally be used for paramagnetic samples. Unpaired electrons occur in the valence-shell, e.g. the d-orbitals of transition-metal ions, or in the uppermost molecular orbitals of paramagnetic molecules (radicals). In either case, there is a series of excited electronic states with energies spread over large ranges, 5000–40 000 cm^{-1} (150–1200 THz). Transitions between these levels are the province of UV or visible spectroscopy. However, if a magnetic field is applied, the ground electronic state can be split into a set of much more closely spaced levels. The spacing depends on the strength of the applied field, but quite modest fields lead to transition energies of a few GHz. As discussed in Chapter 1, these energies lead to high sensitivity, and EPR is the most sensitive of all the techniques considered here.

There is a considerable parallel between EPR and NMR. Both depend on the magnetic moment associated with a spinning particle, either an electron or a nucleus. For a spin of 1/2, the two states produced by interaction with a magnetic field, B, differ in energy by

$$\Delta E = g\mu B$$

where the value of g depends on the identity of the particle, and μ is its magnetic moment. For a free electron, g is 2.0023 while for a proton it is 5.5856. The basic magnetic moments of nuclei and electrons differ considerably because of the dependence on mass, which enters as a reciprocal

$$\mu = eh/4\pi m$$

so that μ_B for the electron (the Bohr magneton) is $9.274 \times 10^{-24} \, \mathrm{J\,T^{-1}}$, while μ_N for the proton (the nuclear magneton) is $5.051 \times 10^{-27} \, \mathrm{J\,T^{-1}}$. It is these factors which are primarily responsible for the difference in characteristic frequencies ($h\nu = \Delta E$) and sensitivities of the two magnetic resonance methods.

EPR spectra depend on determining the g-values for the unpaired electrons in the sample, which differ from those of free electrons and are sensitive to the chemical environment of the paramagnetic atom. This is normally done by using a fixed frequency, usually 9 GHz, and varying the applied field. Note that this is the opposite of the method used in NMR. The reason for this is that, with radiation in the GHz region, with wavelengths of a few millimetres or centimetres, the sample has to be placed in a chamber which is tuned to the particular wavelength used. It is then not possible to vary the wavelength without having an adjustable chamber; in practice it is easier to use a fixed wavelength, and to vary the magnetic field. Quite wide ranges may have to be scanned, since g may be anywhere from just less than 2.0 up to about 10. The frequency of 9 GHz was chosen because systems using it had already been developed from marine radar (X-band), and were well understood. At this frequency, a magnet of about 0.33 T is required. For similar reasons, some spectrometers operate at about 36 GHz (airport radar, Q-band, requiring about 1.4 T), but they are experimentally more difficult to use.

5.1 EXPERIMENTAL CONSIDERATIONS

5.1.1 The sample

Samples for EPR may be liquids, solids, or solutions. In principle, gases may also be examined. Small amounts only are required: sample tubes usually have diameters of a few millimetres, and are packed to depths from a few millimetres to 2–3 cm.

Owing to the very high sensitivity of the technique, it is essential to avoid traces of additional paramagnetic materials. Sample tubes are therefore normally made of silica rather than glass, which may contain iron(III) ions. Similarly, it is sometimes necessary to remove the air from sample tubes containing a powdered solid, or to degas a solution, so that spurious or broadened signals are not obtained from the presence of molecular oxygen. This is particularly necessary for free-radical samples, which usually give sharp spectra with narrow, well-resolved lines. It is rarely required for transition-metal complexes because they often show very broad lines, owing to spin–spin relaxation effects. Such effects occur when the paramagnetic centres are too close together, i.e. in adjacent small molecules, so that the electron spins on one centre can affect those on adjacent centres. It is therefore sometimes advantageous to dilute the sample. If it is soluble, this is simple, and concentrations of about $5 \, \mathrm{mmol\,dm^{-3}}$ or less usually give sharp spectra; however, water is not the best solvent, owing to its high dielectric constant which leads to electric loss, microwave heating and other problems. Solid samples cannot be satisfactorily diluted merely by mixing with an innocuous powder. Dilution must occur at the molecular level, i.e. by recrystallization in the presence of a diamagnetic substance which is isomorphous with the sample, in order to obtain a uniform distribution of the paramagnetic substance within the diamagnetic host.

The use of low temperatures is sometimes advantageous, to increase sensitivity further by slowing down relaxation processes and giving narrower lines. It should be remembered that solutions may well freeze; cooling should therefore be as rapid as possible, to prevent the sample crystallizing out and forming a separate phase. For aqueous solutions, it can be helpful to add another solvent, such as dimethylsulphoxide or glycerol, to ensure glass formulation.

5.1.2   The spectrometer
Spectrometers operating at about 9 GHz (X-band) are usual, but other frequencies are also available, both higher and lower. The chief advantages of a higher frequency are increased resolution and reduction of second-order effects, but there may also be some line sharpening if relaxation processes are occurring with half-lives in the nanosecond range. The higher-frequency spectrometers are rather more difficult to operate, and the X-band is more common. The sweep-range is normally selected so as to cover the likely region of interest, and a marker with a known g-value is often used as a reference. Common reference materials are DPPH (diphenylpicrylhydrazyl radical, $g = 2.0036$), and pitch (2.0028). A spectrum consists of a field-sweep so that, if a signal is observed at lower field than the standard, the corresponding g-value is higher, and the larger g-values occur on the left-hand (low-field) side of the spectrum. The g-value is calculated from the recorded value of the magnetic field at the resonance position:

$$g_{sample} = g_{std} \cdot B_{std} / B_{sample}$$

In order to reduce the noise in the spectrum, most spectrometers employ a modulation technique which superimposes a cyclic variation on the magnetic field. Detection of the signal then involves examining the phase of the resultant wave using a phase-sensitive detector. The practical result of this is that spectra are displayed in first-derivative form, dI/dB, rather than as simple absorption peaks (see Fig. 5.1).

a   b   c

Fig. 5.1 — EPR spectra as (a) intensity versus magnetic field (b) first-derivative form, (c) second-derivative form.

This means that the position of the maximum of absorption is the point at which the central trace crosses the baseline, and the signal width is the distance between the maximum and the minimum (peak-to-peak). Resolution of fine structure can often be improved by taking the second derivative, and some spectrometers are equipped for this. Similarly, some modern spectrometers work in the Fourier-transform mode (FT, see Chapter 1 for corresponding NMR spectrometers); they are able to give considerable improvement in signal:noise ratio.

5.2 FUNDAMENTALS

5.2.1 g-Values

For a free electron in a vacuum, $g = 2.0023$. In chemical systems, the unpaired electron occupies an orbital which may be more or less localized on a single atom or may be heavily delocalized across a molecule or radical. The g-value now reflects the nature of this orbital. In the free-radical case, g remains close to the free-electron value. In other cases there may be a considerable contribution to the magnetic moment of the electron from its orbital motion. The value of g may then be quite different from 2.0; it seldom falls much below this figure, but may reach 9.0 or more. Orbital contributions arise most commonly in systems containing transition-metal ions, and are very similar to those observed in measurements of the magnetic moment.

In principle, each atomic orbital has an associated orbital angular momentum, characterized by the quantum number m_l. For the d-orbitals, an electron in d_{z^2} has zero angular momentum about the z-axis ($m_l = 0$), those in d_{xz} and d_{yz} have one unit, and those in d_{xy} and $d_{x^2-y^2}$ have two units ($m_l = \pm 1, \pm 2$ respectively). The total orbital angular momentum of the atom is related to the quantum number L, which is calculated by summing the m_l-values for all the electrons. For this book-keeping purpose, it is assumed that electrons preferentially occupy the orbitals with highest m_l. Thus, a d^4 configuration would correspond to $L = (+2) + (+1) + (0) + (-1) = 2$. [This procedure is adopted because L is equal to the largest value of M_L, and M_L is the sum of the individual m_l-values: $M_L = L, L-1, \ldots, -(L-1), -L$]. Similarly, the total spin of the system is $S = n/2$, where n is the number of unpaired electrons. For an isolated atom, the total angular momentum, spin plus orbital, is characterized by the quantum number J, which takes all values between $L + S$ and $|L - S|$ and the value of g is given by

$$g = 1 + [J(J+1) + S(S+1) - L(L+1)].[2J(J+1)]^{-1}$$

This formula works well for cases in which there is very little or no influence of surrounding atoms on the orbital momenta of the unpaired electrons, i.e. for free, gaseous atoms and for compounds of the lanthanide elements. In the latter case, the $4f$ electrons are well shielded by the other electrons present.

For the three d-series and the $5f$-elements, the behaviour of the unpaired electrons is considerably influenced by the ligands, with the result that the orbital contribution may be severely reduced or even completely quenched. For instance, in complexes of the $3d$-metals, it is usual to calculate the magnetic moment on a 'spin-only' basis, which often gives quite good agreement with experiment. There is usually some residual orbital contribution, which varies in amount according to the electron configuration and the stereochemistry, and depends on the particular orbitals occupied by the unpaired electrons.

This situation arises because the five d-orbitals do not remain degenerate. In a complex there is a ligand-field splitting of the orbitals into different energy groups. This means that the unpaired electrons do not have unrestricted access to all the orbitals and cannot therefore have their full complement of orbital angular momentum. For instance, it nearly always happens that d_{xy} and $d_{x^2-y^2}$ have very different energies; e.g. in an octahedral complex they are separated by Δ_o which is $8000 \, \text{cm}^{-1}$ or more. These are the orbitals associated with the largest component of angular momentum about the z-axis, $m_l = \pm 2$. In order to display the full momentum, the electron must have equal access to both orbitals with the same value of $|m_l|$. Thus, a large contribution is lost by the lack of degeneracy of these two orbitals. However, d_{xz} and d_{yz} are often degenerate, or very nearly so. Therefore, any configuration which is orbitally degenerate and contains unpaired electrons in these orbitals will show an orbital contribution and will have a g-value different from 2.0. Examples of such configurations are tetrahedral d^3 and octahedral d^7 (high-spin), or d^5 (low-spin) (see Fig. 5.2). (For reasons given below, systems with even numbers of unpaired

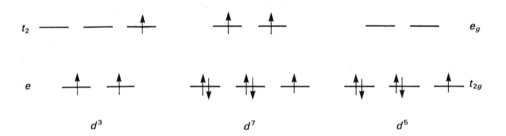

Fig. 5.2 — Orbitally degenerate d^n configurations which give readily observable EPR spectra.

electrons often do not give readily observable EPR spectra.)

Unfortunately, it is not possible to assign a unique value of m_l to each orbital (except when $m_l = 0$). This is because, in solving the wave equation, m_l appears with an imaginary coefficient in terms such as $\exp(im_l\phi)$, where i is $\sqrt{(-1)}$. The 'real' orbitals which chemists use are combinations of the imaginary ones [remember that $\cos \phi = \exp(i\phi) + \exp(-i\phi)$]. The practical result of this is that both d_{xz} and d_{yz} correspond to $m_l = +1$ and to $m_l = -1$, and similarly for the $m_l = \pm 2$ orbtials. Thus, in order for the electron to display the full angular orbital momentum corresponding to m_l, it must have equal access to both orbitals of the pair.

Even in those cases where such a direct orbital contribution is not expected, there can be an indirect effect. When a substance containing unpaired electrons is placed in a magnetic field, the motion of the electrons is affected so that the magnetic moment so induced opposes the applied field (this is the origin of the magnetic susceptibility). The simplest way to describe this effect is in terms of the system gaining some of the characteristics of excited states. It usually happens that the excited states have an orbital momentum and, to the extent that they get 'mixed in' to the ground state description, so the magnetic moment and g depart from the spin-only value. The extent of this mixing is inversely related to the energy of the excited state and there are also numerical factors which depend on the electronic configuration and the extent of covalency in the bonding (see section 3.1.2), but the general form is

$$g = 2.0023(1 - f\lambda/\Delta E) \tag{5.1}$$

where λ is the spin-orbit coupling constant, f is the composite numerical factor, and ΔE is the energy of the state which is being mixed in. Note that λ can be positive or negative, so that g may be greater or less than 2.0023. λ is positive for atoms with fewer than five d-electrons, zero for d^5, and negative for more than five d-electrons. Its absolute numerical value depends on the oxidation state of the metal, but it also increases rapidly with increasing atomic number, so that the effect is relatively large for the later 3d-metals and substantial for the 4d- and 5d-metals.

5.2.2 Fine structure

A species with spin S has total of $2S + 1$ energy states characterized by the quantum numbers M_S. In the absence of a magnetic field all states with $M_S \neq 0$ are expected to remain doubly degenerate (i.e. M_S and $-M_S$ correspond to the same energy). However, these doublets are usually separated from each other by the electric fields produced by the other atoms, which act *via* spin–orbit coupling. The extent of this **zero-field splitting** depends on the structure of the sample, the extent of spin–orbit coupling, etc. In principle it is zero for strict cubic symmetry (Fig. 5.3) but, since such ideal behaviour is rarely seen, zero-field splitting is present in the vast majority of cases. The application of a magnetic field now removes the remaining degeneracy, and transitions may be observed between adjacent states, following the selection rule $\Delta M_S = \pm 1$. In principle, then, $2S$ transitions can occur, their separations representing the extent of zero-field splitting (Fig. 5.4). The appearance of more than one line (when $S > 1/2$) is known as **fine structure** in the spectrum. Whether or not all or any of the transitions can be observed, and what their g-values are, depends on the electron configuration and the zero-field splitting and, of course, on the magnetic-field range of the spectrometer. It should also be mentioned that it is not entirely unknown for 'forbidden' lines to appear, corresponding to $\Delta M_S = \pm 2$.

It frequently happens that the zero-field splitting is very large. Under these circumstances, systems in which n is even may have no resonances within the observable range, and those with odd n show only a single resonance corresponding to the $M_s = -1/2 \rightarrow +1/2$ transition (Fig. 5.4).

5.2.3 Anisotropy

While many systems may approximate reasonably closely to cubic symmetry, many more do not. Such systems are called *anisotropic*. Under these conditions, the value of g depends on the direction between the magnetic field and the principal (z) axis of the system. It often happens that the symmetry is axial, i.e. the x- and y-directions are equivalent but different from z. This will be the case when the principal axis is a

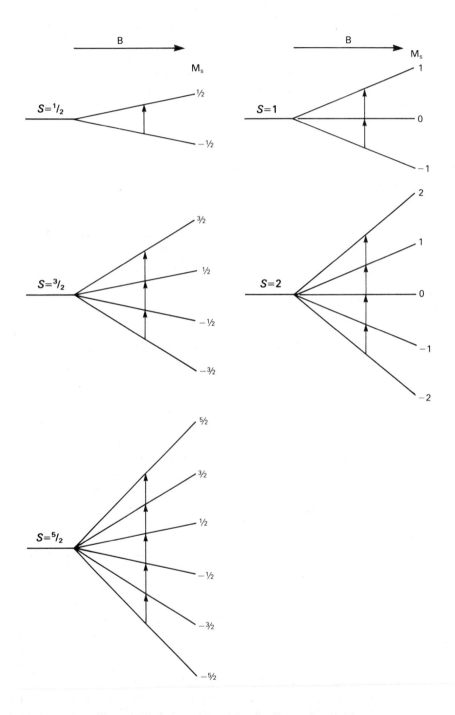

Fig. 5.3 — Energy-level diagrams for various values of S and no zero-field splitting. Note that the applied magnetic field increases from left to right, and that all transitions for any one system occur at the same field-value.

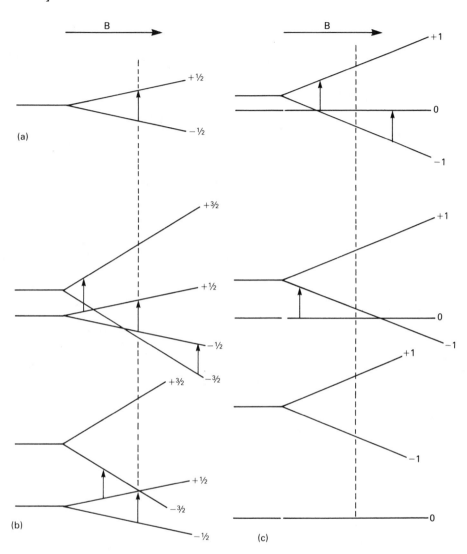

Fig. 5.4 — Energy-level diagrams: (a) $S = 1/2$; (b) $S = 3/2$ for two different zero-field splittings; (c) $S = 1$ for three different zero-field splittings. The dotted vertical lines mark the position of the (single) absorption line in the absence of zero-field splitting.

symmetry axis of order three or greater. For instance, many complexes of copper(II) have tetragonal symmetry. There are then two principal g-values, $g_{\parallel}$ and $g_{\perp}$; when the field and molecular axes are parallel, the signal appears at $g_{\parallel}$, when they are perpendicular it appears at $g_{\perp}$. (Strictly speaking it is the axis of the spin angular momentum which applies rather than the molecular symmetry axis; these are often the same, but may differ.) If the sample is a single crystal in which all the molecules are oriented similarly with their axes at an angle θ to the magnetic field, the observed g-value is given by

$$g^2 = g_{\parallel}^2 \cos^2\theta + g_{\perp}^2 \sin^2\theta \; .$$

If the symmetry is lower still, three principal g-values are needed and the observed g-value for an arbitrary orientation is

$$g^2 = g_x^2 \cos^2\theta_x + g_y^2 \cos^2\theta_y + g_z^2 \cos^2\theta_z$$

where the angles are those between the field and the molecular axes. The geometric terms are all less than unity, so that the principal g-values are the extremes between which the observed values must lie. If the crystal is rotated in the magnetic field, the g-value moves through a range defined by the largest and smallest of the principal values. There is also a small dependence of the signal-intensity on orientation.

The origin of the anisotropy effect lies in the mixing-in of excited states referred to above [eq. (5.1)], since the states which can be mixed depend on the relative orientation of the spin and magnetic-field axes. For a d^1 system in an axially symmetric situation, such as for VO^{2+}, the d-orbital splitting will be as shown in Fig. 5.5(a). In the ground state, the single electron occupies the d_{xy}-orbital, and

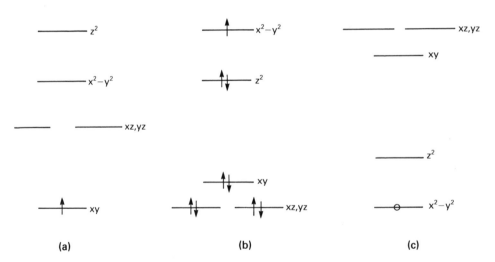

(a)   (b)   (c)

Fig. 5.5 — d-orbital splitting diagrams for (a) VO_2^+ (d^1), (b) Cu^{2+} (d^9) in tetragonal geometry, and (c) Cu^{2+} in the 'hole' formalism.

there are excited states corresponding to the occupation of the other orbitals. $g_{\parallel}$ now depends (inversely) on the energy of $d_{x^2-y^2}$, since this is the orbital which has angular momentum about the z-axis (the parallel direction); similarly, $g_{\perp}$ depends on the energy of $d_{xz,yz}$ (which have angular momentum about the y- and x axes equal to that of d_{xy} about these axes). Since λ is positive for d^1, both g-values will be less then the free-electron value and $g_{\parallel}$ will be smaller than $g_{\perp}$. Similar arguments apply to copper(II) (Fig. 5.5(b)); this ion, with a d^9 configuration, can be treated as a 'one unpaired hole' case, the hole occupying $d_{x^2-y^2}$

[Fig. 5.5(c)]; the two relevant excited states have the hole in d_{xy} ($g_\parallel$) or $d_{xz,yz}$ ($g_\perp$). (d_{z^2} is not considered because it has no angular momentum.) Since λ is now negative, $g_\parallel$ is greater than $g_\perp$.

In a solution, the molecules are continuously tumbling, usually at rates considerably greater than the frequency of the spectrometer. Any anisotropy is then averaged to zero during the time required to excite the system. Thus, even though at any one instant the collection of molecules has all possible orientations relative to the magnetic field, they will appear to be isotropic and give the same g-value.

When the solution is frozen, or if the sample is a powdered solid, all possible orientations still occur randomly, but the molecular positions are fixed. Each molecule with a particular orientation has its own g-value and the observed spectrum is the sum over all the molecules. On a purely statistical basis, there will be very few molecules with principal axes parallel to the field, and far more perpendicular. This factor combines with angle-dependence of intensity to give the distribution of intensities shown in Fig. 5.6. In the derivative spectrum it is usually possible to

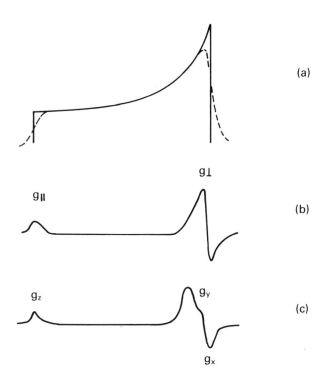

Fig. 5.6 — Typical shapes of EPR spectra for powders or frozen solutions containing anistropic species with $g_\parallel > g_\perp$. (a) and (b) are for an axial system (tetragonal) in the intensity and first-derivative form, (c) is the first-derivative form for a rhombic system.

recognize the two principal g-values, though the resolution depends critically on the difference between $g_{\parallel}$ and $g_{\perp}$. For many rhombic systems ($x \neq y \neq z$), the difference between g_x and g_y is not great, and the spectrum appears to have an asymmetry in the '$g_{\perp}$' line. If the three principal g-values are appreciably different, three major lines appear in the spectrum.

5.2.4 Hyperfine structure
An EPR spectrum may show additional fine structure when the atom on which the unpaired spin is centred has a nucleus which also has a spin. In a manner exactly analogous to that in NMR spectroscopy, a nucleus of spin I gives rise to a splitting of the EPR line into $2I + 1$ components, all of equal intensity separated by the coupling constant, A. That is, in the presence of the magnetic field, the nucleus can adopt any one of $2I + 1$ orientations, each of which produces a different local magnetic field at the electron. Each energy level is thus split into $2I + 1$ components for any given value of the applied field, and the energy depends on the product of M_S and m_I, resulting in $2I + 1$ lines in the EPR spectrum (Fig. 5.7). This is known as **hyperfine structure**.

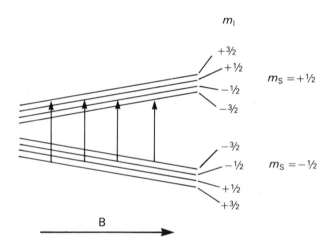

Fig. 5.7 — Hyperfine splitting of the $m_s = -1/2 \rightarrow +1/2$ transition due to interaction with a nuclear spin $I = 3/2$. The four resulting transitions are equally spaced.

In some cases, the unpaired electrons are delocalized over several identical atoms, e.g. in polynuclear metal complexes or organic radicals. The splitting is then into $2nI + 1$ lines (n = the number of equivalent nuclei).

The value of the hyperfine coupling constant, A, depends on the amount of interaction which occurs between the unpaired electron and the nucleus. There are three major factors which control the magnitude of A. First, the orbital containing the unpaired electron may have some s-character. To this extent it is able to enter the nucleus directly, since s-electrons have finite probability of occurring at the origin. This is the Fermi-contact term, which gives a positive contribution to A. For transition-metal systems, this implies a relatively low symmetry for the system, since the unpaired electrons are formally d-electrons and mixing of d- and s-orbitals can only occur when they have the same symmetry classification.

Secondly, there are spin-polarization (exchange) effects. The presence of unpaired spin density in the valence orbitals polarizes the distribution of electrons in the core orbitals. Electrons in other orbitals with spin parallel to that of the unpaired electron will spend more time in its vicinity; consequently, the distribution of spin in the core orbitals is not uniform. However, the distribution of spin in the core s-orbitals is also polarized by the nuclear spin (see Fermi contact in Chapter 2). Since two separate spin systems are acting on the same core electrons, the core electrons effectively couple the spin systems together. For s-orbitals which lie inside the d-orbital concerned (i.e. $1s$ and $2s$, for unpaired spin in $3d$-orbitals), the electronic spin at the nucleus is opposite in sign to that of the unpaired electrons. The coupling is therefore negative. For the s orbital which is concentrated beyond the d-orbital, the coupling is positive.

Finally, there is a coupling between the magnetic moments of the nucleus and the unpaired electrons, just as if they were two small physical magnets. Such dipolar coupling depends not only on the distance between the two magnets but also on their relative orientations. The orientation of the nuclear spin is governed by the applied field, while that of the unpaired electron depends on which orbital it occupies and on the orientation of the molecule relative to the applied field. This effect is therefore anisotropic; there are two (or three) principal values, $A_{\parallel}$ and $A_{\perp}$ (or A_z, A_y, A_x), which may be positive or negative.

The values of all these components also depend on the extent of covalency in the metal–ligand bonds, since this will result in some delocalization of the electron away from the nucleus with which it is being coupled. In general, increased covalency leads to diminution in the numerical value of A.

The observed A-value is the sum of all these contributions, and its magnitude depends on the relative magnitudes and signs of the various components:

$$A_{obs} = A_{iso} + A_{aniso}$$

where A_{iso} is the sum of the Fermi-contact and polarization terms, and A_{aniso} is the relevant anisotropic term for the particular orientation concerned. If the system is anisotripic, the coupling constant will obviously vary with the orientation of the molecule. Thus, for a single crystal, A will change with orientation, just like g, with extreme values of $A_{\parallel}$ and $A_{\perp}$. For powders and frozen solutions all orientations are possible, but the principal A-values will be associated with the corresponding g-value. Thus, for an axial system, the $g_{\parallel}$ line will appear to be split into $2I+1$ components separated by $A_{\parallel}$, and the $g_{\perp}$ line will be similarly split with separation $A_{\perp}$. The full structure of the spectrum may not be resolved; sometimes all that can be seen is a broadening of the lines. For solutions, the anisotropic effects are averaged to zero by the rapid tumbling of the molecules, and only A_{iso} remains. It therefore follows that

$$A_{iso} = (A_{\parallel} + 2A_{\perp})/3 \ .$$

In practice there may be discrepancies of 10–20% between this average and the true value of A_{iso}.

It is often found that the inner lines of a hyperfine-split pattern are more intense than those of the outer lines, and the separation of the lines may not be uniform (see Fig. 5.8). These are 'second-order' effects, and are more likely to appear with

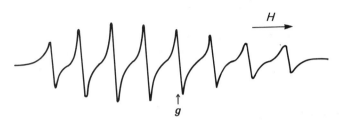

Fig. 5.8 — EPR spectrum of VO^2 in aqueous solution. Note the non-uniform spacing and intensities. (Reproduced with permission from B. A. Goodman and J. B. Raynor, *Adv. Inorg. Chem. Radiochem.*, **13** (1970) 223.)

spectrometers operating at relatively low frequencies. In fact the lines should all have the same total intensity (i.e. integrated area) but there is often some broadening of the outer lines so that their height is diminished. The possible causes of this effect are small degrees of anisotropy and insufficiently rapid tumbling of molecules in solution. More detailed dicussions of these features can be found, for example, in Symons's book (see Bibliography). The variation in height is never so great as to cause confusion with splitting patterns due to several nuclei (see below and Problem 5.6).

The spacing of the hyperfine lines sometimes increases or decreases progessively across the pattern (*cf.* Fig. 5.8). The value of the hyperfine coupling constant, A, can then be obtained by averaging the separations of the outermost pairs of lines. The principal problem is that the apparent centre of the pattern no longer corresponds exactly to the true g-value. Discrepancies of the order of a few percent may occur, which might be misleading if comparison is to be made with data from the literature. Uneven spacing of the hyperfine lines is also sometimes due to coupling with the nuclear quadrupole moment. Only nuclei with $I > 1/2$ possess quadrupole moments, and the effect is normally seen for nuclei with relatively large moments and when there is a substantial electric-field gradient present at the nucleus (i.e. a low-symmetry site — see Chapters 3 and 4).

5.2.5 Superhyperfine structure
In an analogous way, nuclei on adjacent atoms may give splitting of the spectrum. The strength of the coupling depends on the extent of delocalization of the unpaired electron on to these atoms, and decreases rapidly with the number of bonds involved; it is usually detectable only for nuclei of atoms directly bound to that containing the unpaired electron. In transition-metal complexes, this means that only the donor atoms of the ligands need be considered. Coupling of this type is classed as **superhyperfine structure**.

The pattern observed is again completely analogous to that seen in NMR, for the same reasons. Thus, n nuclei of spin 1/2 give $n + 1$ lines with intensities in the ratios of the binomial coefficients (see section 2.2.2), and n nuclei of spin I give splitting into $2nI + 1$ lines whose intensity patterns can be derived by the J-tree technique. In a first-derivative spectrum, the number of lines present can be determined by counting the number of peak maxima (or minima); this is usually simpler than trying to identify each complete derivative curve.

Every pattern is again characterized by a coupling constant which reflects the extent of delocalization of the unpaired electron on to the atoms concerned. It is therefore a measure of, for instance, the covalency of a metal–ligand bond. As for hyperfine coupling, the A-value can be anisotropic.

5.2.6 Line widths
The theoretical width of an EPR absorption line is very small, about $0.1\,G$ ($10\,\mu T$). Lines as narrow as this are often found for solutions of organic free radicals. However, as indicated in Chapter 1, various relaxation processes may shorten the lifetime of the excited state and lead to line broadening. This is particularly important for transition-metal complexes, where spin–lattice and spin–spin relaxation may occur. In particular, if there are any low-lying excited electronic states relaxation becomes quite efficient, and lines are considerably broadened. Examples are given below, but this is a further complication for even-electron systems.

5.3 INTERPRETATION
EPR spectra are usually obtained as plots of the first derivative of absorption intensity versus applied magnetic field. g-values are obtained by comparing the field value at the signal position with that of the standard:

$$g = g_{std} \cdot B_{std}/B_{obs}$$

Similarly, line-spacing in coupling patterns is measured first in terms of differences in magnetic field, and is often quoted in units of magnetic field rather than hertz. (For interconversion of units, $10\,G = 1.0\,mT \equiv 0.935 \times 10^{-3}\,cm^{-1}$, when $g = 2.00$; otherwise, $h\nu = 46.69 g\Delta B\ T^{-1}m^{-1}$).)

A large number of factors affect the shape, position and width of an EPR spectrum. It is relatively easy to extract approximate parameters, i.e. the principal g- and A-values, but second-order effects often make it difficult to obtain accurate values. For instance, hyperfine splitting patterns may not be symmetrical, and the line-spacing may change across the pattern. Usually the change is systematic, and the A-value may be estimated reasonably reliably as the average of the outermost splittings. Under these circumstances, the g-value does not correspond to the magnetic field at the centre of the pattern (see Fig. 5.8). If accurate values are required, it is necessary to compare the experimental spectrum with one simulated with trial values of the parameters, and to adjust the parameters to obtain the best correspondence. It often happens that two (or more) parameters have rather similar effects on the shape of the spectrum, and intelligent guesswork may have to be employed. What this really means is that the parameters are correlated (in terms of the spectrum shape) and none can be determined to high precision.

Similarly, relative intensities can be approximated by comparing the peak-to-peak (maximum-to-minimum) separations on the first-derivative display, when more than one species is present and the spectra are relatively simple and there are no significant differences in line width. Separate samples can be compared by comparison with the same reference standard. For most accurate work, or if the spectra are at all complicated, double integration of the first-derivative spectrum is required.

A number of simple qualitative deductions can be made readily, e.g. identification by 'fingerprinting' (i.e. comparison of the experimental parameters with those already known), deduction of the number of species present (if the spectra are relatively simple), identification of which metal ion in a mixture has undergone oxidation or reduction (particularly if nuclei with $I > 0$ are present, which can be recognized from their hyperfine patterns). In most real-life cases it is necessary to have some background knowledge about the particular system, and illustrative examples are discussed below. More detailed treatments can be found in the references cited in the Bibliography.

5.3.1 Single-electron systems

5.3.1.1 Radicals

Free-radicals, whether organic or inorganic, usually give sharp spectra with g close to 2.00. Hyperfine and superhyperfine structure may be seen, and is normally well-resolved. One of the simplest inorganic radicals is also one of the oldest known: Fremy's salt, the potassium salt of the anion $(O_3S)_2NO^{2-}$. Dilute solid solutions of this in the lattice of the diamagnetic dimeric form can be obtained relatively easily. A sharp signal is observed, centred at $g = 2.0055$ which shows well-resolved hyperfine splitting into three components by the nitrogen nucleus (^{14}N, $I = 1$), $A = 1.309$ mT. The neighbour atoms, ^{16}O and ^{32}S have spin of zero, so there is no superhyperfine splitting. Similarly, reaction of $S(=NBu^t)_2$ with Me_2SiCl_2 in the presence of lithium gives a cyclic molecule, which was identified by the EPR spectrum of its radical cation, **5.I**, obtained by chemical oxidation (Fig. 5.9). The spectrum shows a 1:2:3:2:1 quintet pattern due to interaction of the unpaired electron with two nitrogen nuclei. When the starting material reacts simply with an alkali metal, a radical anion is produced, whose spectrum is shown in Fig. 5.10; in this case, additional fine structure can be seen due to coupling with the hydrogen atoms $[A_{iso}(^1H) = 0.024$ mT].

$$
\begin{array}{c}
Bu^t \\
| \\
N \\
\diagup \quad \diagdown \\
S \qquad \qquad SiMe_2 \\
\diagdown \quad \diagup \\
N \\
| \\
Bu^t
\end{array}
\Biggr]^{\overset{+}{\cdot}}
$$

5.I

When several magnetic nuclei are present, the spectra can become quite complicated. For instance, Fig. 5.11 shows the EPR spectrum of the radical obtained by treating $(MeO)_3P \rightarrow BH_3$ with Bu^tOOBu^t under UV irradiation. The likely

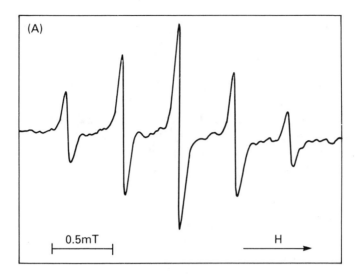

Fig. 5.9 — EPR spectrum of compound **5.I** ($g = 2.0059$, $A(^{14}N) = 0.45$ mT; satellites due to ^{29}Si can just be seen, $A = 0.600$ mT). (Reproduced with permission from W. Kaim, *J. Chem. Soc., Chem. Comm.*, (1989) 470.)

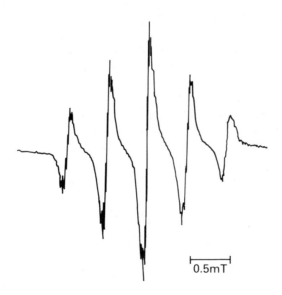

Fig. 5.10 — Solution EPR spectrum of $S(=NBu^t)_2^-$ ($g = 2.0071$, $A(N) = 0.515$ mT). (Reproduced with permission from J. A. Hunter, B. King, W. E. Lindsells, and M. A. Neish, *J. Chem. Soc., Dalton Trans.*, (1980) 880.

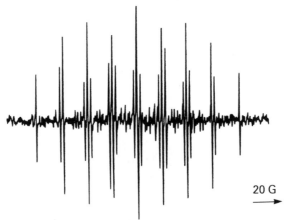

20 G

Fig. 5.11 — EPR spectrum of $(MeO)_3PBH_2^{\cdot}$. (Reproduced with permission from J. A. Baban and B. P. Roberts, *J. Chem. Soc., Perkin Trans. II*, (1984) 1717.)

product is that formed by hydrogen atom abstraction, $(MeO)_3PBH_2^{\cdot}$. This is a good model on which to illustrate the method of analysis of complex spectra, since 21 major lines can be seen. These arise because three types of magnetic nucleus are present: ^{31}P, $I = 1/2$, ^1H, $I = 1/2$, ^{11}B, $I = 3/2$ (natural boron also contains 18.8% of ^{10}B, $I = 3/2$, which gives extra sets of lines of relatively low intensity; in initial analysis, these may be ignored). It is necessary first to identify the coupling constants and line-groupings. The splitting of the outermost pairs of lines (14.6 G = 1.46 mT) is likely to be one of the coupling constants. Since lines 2 and 20 have the same intensity as lines 1 and 21, and several other lines of the same intensity occur at the same spacing (lines 4, 6, 7, 9, etc.), this is likely to be the ^{11}B coupling, which give splitting into sets of four equal lines. On this basis, lines 1, 2, 4 and 7 belong to one group of four. The separation between lines 1 and 3 (16.6 G) is likely to represent another coupling constant, and since they are in intensity ratio 1:2, they probably form part of a 1:2:1 triplet, from coupling with two ^1H nuclei; this is confirmed by the presence of another line in the expected place (line 6). The final coupling constant, $A(^{31}P)$, can be identified by recognizing that the separation between the outermost lines (1 and 21) represents the sum of the coupling constants to all the individual nuclei, with each weighted by twice the nuclear spin: $\Sigma_i 2I_i A_i$. In this case, the overall separation is 121 G, which leads to:

$$2(3/2)14.6\,G + 2(1/2)16.6\,G + 2(1/2)16.6\,G + A(^{31}P) = 121\,G$$
$$^{11}B \qquad\qquad ^1H \qquad\qquad ^1H$$

Hence, $A(^{31}P) = 44.0\,G$. (Coupling to the hydrogen atoms of the methyl groups is expected to be too small to be resolved.) The complete coupling pattern may now be constructed, using the same J-tree technique as for NMR (Fig. 5.12), and final confirmation obtained by calculating a full simulation of the spectrum.

There has been much study of radicals obtained by gamma-irradiation of transition-metal compounds, which is a very convenient way of obtaining sufficient

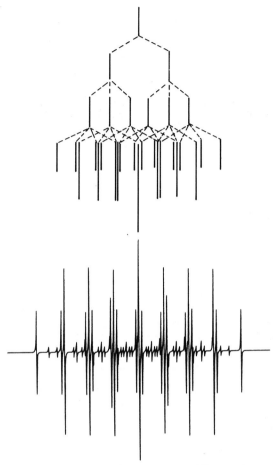

Fig. 5.12 — J-tree for $(MeO)_3PBH_2^{\bullet}$, showing coupling to ^{31}P, two ^1H and ^{11}B, and the full simulated spectrum (the latter reproduced with permission from J. A. Baban and B. P. Roberts, *J. Chem. Soc., Perkin Trans. II* (1984) 1717).

dilution to avoid relaxation effects. In these cases, the principles of spectrum-analysis are the same as for conventional paramagnetic transition-metal compounds, to which the remainder of section 5.3.1 will be devoted.

5.3.1.2 d^1 systems

In strict octahedral geometry, spin–orbit coupling gives a ground state in which the orbital and spin angular momenta exactly cancel. At very low temperatures, therefore, the system would appear diamagnetic and would show no EPR spectrum. At normal temperatures both paramagnetism and an EPR signal are expected, with an orbital contribution which may result in g-values considerably less than 2.0. However, in practice the geometry is never strictly octahedral. Small distortions occcur which remove the triple orbital degeneracy and give a set of closely spaced levels. These are the conditions required for rapid spin–lattice relaxation; the spectra are severely broadened, and can only be observed at low temperatures.

When distortion from cubic symmetry is very large, EPR spectra can be obtained readily. The most widely studied example is VO^{2+}, which forms six-coordinate complexes with tetragonal symmetry (or lower) (see Chasteen, Bibliography). There will now be two (or three) principal g-values, the magnitudes of which depend on the indirect orbital contributions from the spin-orbit coupled mixing-in of excited states (see section 5.2.2):

$$g = 2.0023 + kn\lambda/\Delta E$$

ΔE is the energy separation between the two states being mixed, which can be measured from the electronic spectrum. The factor k is always less than unity; it takes account of covalency which delocalizes the unpaired electron, and reduces its orbital contribution; its value is related to the coefficients of the d-orbitals in the molecular orbitals. The remaining factor, n depends on the identity of the two orbitals which are being mixed, and can be determined from the 'magic pentagon' (Fig. 5.13). Thus,

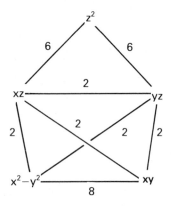

Fig. 5.13 — The 'magic pentagon', showing the coefficients for mixing of pairs of d-orbitals.

in the case of VO^{2+}

$$g_{\parallel} = 2.0023 - 8k\lambda/\Delta E_{x^2-y^2}$$

$$g_{\perp} = 2.0023 - 2k'\lambda/\Delta E_{xz,yz}$$

Since λ is positive, both gs are less than the free-electron value: typical ranges are $1.97 > g_{\parallel} > 1.93$ and $2.0 > g_{\perp} > 1.96$. Many measurements are made on solutions, when only the average g-value is obtained; this is usually reported as g_{iso}.

Vanadium-51 (99.8%) has a nuclear spin of 7/2 and hyperfine structure is usually seen. The coupling constant is anisotropic, with $A_{\parallel}$ always considerably greater than $A_{\perp}$; this means that the eight components of the parallel signal straddle the more closely spaced perpendicular peaks. A typical spectrum is shown in Fig. 5.14. Typical

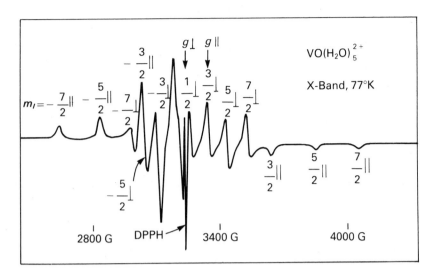

Fig. 5.14 — EPR spectrum of VO_2^+ in frozen aqueous solution, showing overlapping of the hyperfine-split $g_{\parallel}$ and $g_{\perp}$ patterns. (Reproduced with permission from N. D. Chasteen, *Biol. Mag. Res.*, **3** (1981) 67.)

ranges are $(130 - 220) \times 10^{-4} \text{cm}^{-1}$ for $A_{\parallel}$ and $(45 - 90) \times 10^{-4} \text{cm}^{-4}$ for $A_{\perp}$. In magnetic-field units these ranges are about 12–20 mT and 4–8 mT, respectively. For solution measurements, only the average value, A_{iso} is seen.

Both A and g reflect the nature of the ligands bound to vanadium, and there are quite good negative correlations between the two parameters. With ligands which are increasingly higher in the spectrochemical series, the d-orbital splittings, ΔE, become greater and g increases. At the same time, the bonding is becoming more covalent and delocalization of the unpaired spin reduces the hyperfine coupling constant. Provided that reasonably similar compounds are compared (e.g. chelates with the same ring size), this provides a method of identifying likely ligand sets (see Fig. 5.15). Because the range of values is greater, best diagnoses are made by comparing $g_{\parallel}$ and $A_{\parallel}$; if only solution data are available, g_{iso} shows much more variation than A_{iso}. Analogous correlations are found for complexes of other metals with d^1 configurations.

5.3.1.3 d^9 systems

The most common d^9 case is copper(II), which has been widely studied. The majority of complexes have distorted six-coordinate structures (Jahn–Teller distortion) in which two bonds are longer than the remaining four. In some cases the distortion is so severe that square-planar co-ordination results. The unpaired electron is in the $d_{x^2-y^2}$-orbital and all other orbitals are doubly occupied. It is also possible for copper(II) to occur in 'tetrahedral' co-ordination, but distortions towards square-planar geometry are again found and the unpaired electron still occupies $d_{x^2-y^2}$. The energy of the d_{z^2}-orbital depends on the degree of distortion from strict octahedral/tetrahedral geometry. When it is relatively close to $d_{x^2-y^2}$ in

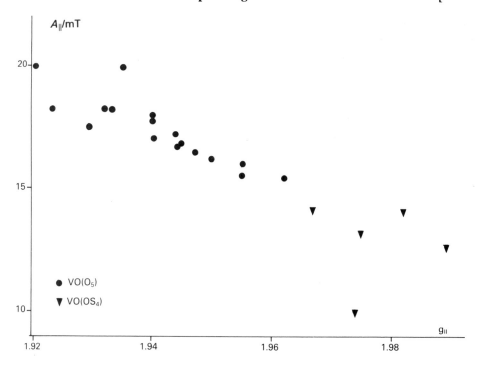

Fig. 5.15 — Correlation between $g_\parallel$ and $A_\parallel$ for complexes of VO_2^+ for ligands with two different sets of donor atoms.

energy (small distortion) there is usually some line-broadening owing to spin-lattice relaxation. No augmentation of g results, because the two orbitals cannot be coupled by spin–orbit effects, having different values of angular momentum about all axes. The line broadening is often sufficient to obscure any quadrupole interaction with the copper nuclei ($I = 3/2$), and may obscure the differential splitting owing to the two isotopes (see below).

The d^9 configuration is conveniently treated by the 'hole' analogy, in which the single positive 'hole' in $d_{x^2-y^2}$ is treated as the spinning entity. Since a hole is effectively a positive charge, this has the result of inverting the orbital energies, to give the pattern shown in Fig. 5.5(c). Spin–orbit coupling can mix the ground state with excited states in which the hole is in d_{xy} or $d_{xz,yz}$, giving

$$g_\parallel = 2.0023 - 8k''\lambda/\Delta E_{xy}$$

$$g_\perp = 2.0023 - 2k'''\lambda/\Delta E_{xz,yz}$$

The spin-orbit coupling constant is negative, and about three times the size of that for vanadium(IV), so that the g values are significantly larger than 2.0. Typical ranges are $2.1 < g_\parallel < 2.35$ and $2.02 < g_\perp < 2.07$.

Hyperfine coupling also occurs. Again, it is expected to be anisotropic, and $A_\parallel$ is usually greater than $A_\perp$ by a factor of about seven†. However, for copper, there is a further complication because there are two isotopes with high abundance, both of which have the same nuclear spin ($I = 3/2$) but rather different abundances and slightly different magnetic moments: ^{63}Cu, 69.1%, 2.22 μ_N, and ^{65}Cu, 30.9%, 2.38 μ_N. Thus, four-line splitting is expected for both isotopes; because of the similarity in magnetic moments, the lines fall close together. The relative intensities are about 2:1. In practice, best resolution is obtained in the lowest-field part of the parallel signal but, even so, the weaker ^{65}Cu signals are often not seen (see Fig. 5.16). Data quoted usually refer to ^{63}Cu.

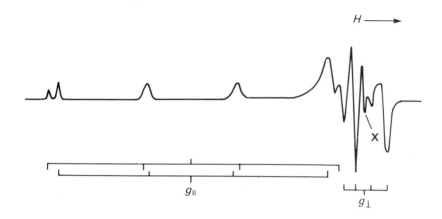

Fig. 5.16 — Frozen-solution spectrum of $Cu(acac)_2$, showing the positions of the $g_{\parallel}$ lines for the two isotopes of copper. For $g_\perp$ the two sets of lines cannot be resolved. Line X is a 'forbidden' transition. (Reproduced with permission from B. A. Goodman and J. B. Raynor, *Adv. Inorg. Chem. Radiochem.*, **13** (1970) 323.)

The two parameters g and A are to some extent correlated. Within related series of compounds, increasing covalency of the metal–ligand bonds leads to decrease in g and increase in A. As the covalencey increases, the energies of the excited states rise, so that the orbital contribution to g becomes less effective, and g becomes closer to the free-electron value. The effect of covalency on the A-values is impossible to predict, since there are so many factors involved, but theory also shows that there is a considerable dependence of A on g. The best discrimination is obtained from the parallel values, since these show the widest variation (Fig. 5.17). Many square planar compounds fall in the region shown on Fig. 5.17, with the complexes of softer ligands lying towards the upper part of the band. Distortion towards tetrahedral geometry gives parameters which lie below this band.

Under good conditions, superhyperfine coupling to ligand donor atoms can be seen. For instance, Fig. 5.18 shows the spectrum of the complex formed between

† Whenever A-values are used here the sign is ignored, i.e. only the magnitude, $|A|$, is given. This is for the practical reason that the sign of A cannot readily be determined. In many cases, $A_{\parallel}$ and A_{iso} are negative.

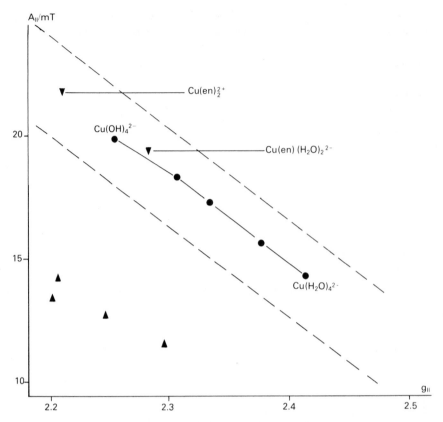

Fig. 5.17 — Correlation between $g_\parallel$ and $A_\parallel$ for complexes of Cu^{2+}. The region between the dashed lines is where data for the majority of square-planar and six-coordinate complexes fall. The three points between $Cu(H_2O)_4^{2+}$ and $Cu(OH)_4^{2-}$ are for the complexes $Cu((LH_2)(H_2O))$, $Cu(LH)(LH_2)^-$, and $Cu(LH)_2^{2-}$, where H_2L is 2,6-dihydroxybenzoic acid. The points in the lower left-hand corner are for distorted tetrahedral complexes of pyridine-type ligands, $CuCl_2(N)_2$.

Fig. 5.18 — Aqueous-solution spectrum of the complex formed between $^{63}Cu^{2+}$ and D-penicillamine at pH 10. The peak marked $\bigcirc$ is attributed to the presence of some ^{65}Cu, and that marked $\times$ is due to an isomer of the principal species. (Reproduced with permission from F. J. Davis, B. C. Gilbert, R. O. C. Norman, and M. C. R. Symons, *J. Chem. Soc.*, *Perkin Trans. II* (1983) 1764.)

Cu^{2+} and D-penicillamine, $HSCMe_2(H_2N)CHCO_2H$. Owing to the change in line width across the spectrum, the hyperfine coupling is resolved only on the highest-field component and on the parallel signal, but both show quintet structure indicating the presence of two nitrogen donor atoms. Comparison of the g_{iso} and $A(^{63}Cu)_{iso}$ values with data for similar complexes of other amino acids and related ligands showed that the second donor atom in the D-penicillamine complexes was the thiolate sulphur atom and not the carboxylate (Fig. 5.19).

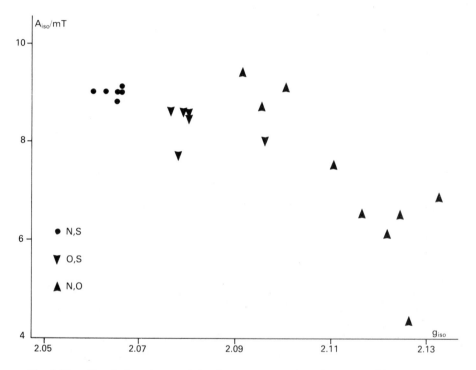

Fig. 5.19 — Correlation of g_{iso} and A_{iso} for square-planar complexes of Cu^{2+} with various ligand donor-atom sets.

A clearer example of superhyperfine splitting is shown in Fig. 5.20, which is the spectrum of $Cd[Cd(SCN)_4]$ which has been crystallized in the presence of a small amount of Cu^{2+} (about 0.2 at.%). The superhyperfine coupling pattern clearly indicates that the copper cation is bonded to the nitrogen atoms of two thiocyanate anions (each with $I = 1$, giving five-line superhyperfine structure). Since Cu^{2+} is substituting for Cd^{2+}, this result implies that the cadmium ions in the anion are bonded to two sulphur atoms and two nitrogen atoms, i.e. that the thiocyanate ligands are statistically distributed. On the other hand, when Cu^{2+} is doped into $Cd[Hg(SCN)_4]$, the superhyperfine splitting indicates four nitrogen ligands for Cu^{2+}; therefore, Hg^{2+} carries four S-bonded thiocyanate ligands, in keeping with the 'soft' nature of mercury(II).

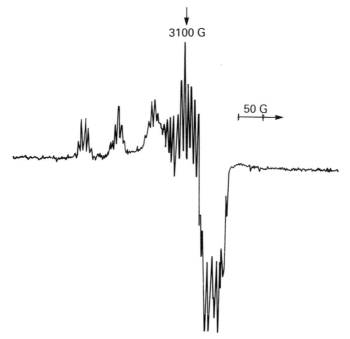

Fig. 5.20 — EPR spectrum of Cu^{2+} doped into $Cd[Cd(SCN)_4]$. (Reproduced with permission from M. C. R. Symons, D. X. West, and J. G. Wilkinson, *J. Chem. Soc., Dalton Trans.* (1975) 1698.)

5.3.1.4 d^5 systems (low-spin)

A low-spin octahedral complex would have a t_{2g}^5 configuration in which the triple degeneracy is expected to be slightly lifted by small distortions. Considerable mixing of the resulting closely spaced energy states will be possible, leading to rapid spin-lattice relaxation and very broad lines. Spectra are only easily observed when the symmetry is considerably lowered; this usually requires the presence of at least two different ligands, and results in the appearance of two or three g-values for powders and frozen solutions. The g-values are, however, usually close to 2.0. Hyperfine splitting due to nuclear spin is usually seen, e.g. for ^{55}Mn ($I = 5/2$), ^{99}Tc ($I = 9/2$), but superhyperfine splitting frequently does not appear.

5.3.2 Multi-electron systems

As indicated earlier, the EPR spectra of systems containing more than one unpaired electron can be difficult to observe or to interpret. In particular, transition-metal complexes with even-electron configurations are very difficult to study. The combined effects of zero-field splitting and rapid relaxation may mean that a spectrum will be so broad and weak that it cannot be observed even at the temperature of liquid helium.

This situation can sometimes be turned to advantage. For example, oxidation of a manganese(II) complex could give either manganese(III) (d^4) or manganese(IV) (d^3); the former does not usually give an observable spectrum whereas the latter is readily detectable. However, negative evidence is always rather dangerous, and it is necessary to remember that powders may not be magnetically dilute, and that there may be direct interactions between metal ions, both of which may lead to broadening and weakening of a manganese(IV) spectrum.

5.3.2.1 d^5 systems (high-spin)

Common examples with high-spin d^5 configurations are Mn^{2+} and Fe^{3+}. These are very common impurities, particularly in salts of metals close to iron in the Periodic Table. Care should be taken over possible spurious peaks, especially if the species under investigation may actually be EPR silent.

The d^5 configuration is unique in that it is an orbital singlet state (i.e. there is only one distinguishable way of distributing the electrons between the orbitals), and there are no excited states with the same spin multiplicity. There is thus no orbital contribution in the ground state and no mixing in of excited states is possible while the symmetry remains high. These are also conditions under which zero-field splitting is small. Hence, in the absence of hyperfine splitting, only a single line is seen, g is isotropic and has a value close to 2.00. Hyperfine coupling is usually seen for Mn^{2+} ($I = 5/2$), giving six-line splitting; A is also isotropic and usually lies in the range 5–10 mT (50–100 G).

If the symmetry is lowered, zero-field splitting may become much larger than the spectrometer frequency, with the result that effectivey only the $M_s = 1/2 \leftrightarrow -1/2$ transition lies within the range normally scanned. For axial symmetry the system is now anisotropic and, while $g_{\parallel}$ remains close to 2.0, $g_{\perp}$ becomes close to 6.0. Powders and frozen solutions therefore show two major resonances with these g-values.

Finally, if the environment of the metal ion has low symmetry (e.g. C_{2v}, D_{2h}) spin–orbit coupling can mix in some of the excited $S = 3/2$ states, resulting in a g-value of about 4.3. Other lines are usually off-scale. This is found, for instance, when the metal ion is in four-coordination of the MA_2B_2 type.

5.3.2.2 d^3 systems

In octahedral geometry, which is strongly favoured for d^3, the ground state is an orbital singlet and other states are high in energy. EPR spectra are therefore sharp and easy to observe, and g is close to 2.0 (slightly below due to spin–orbit coupling effects). Electrons in t_{2g} orbitals can interact only with π-bonding ligands, and superhyperfine coupling is rarely seen.

5.3.2.3 d^7 systems (high-spin)

The commonest example of a high-spin d^7 ion is Co^{2+}. It has three unpaired electrons and therefore behaves somewhat like the d^3 case. Zero-field splitting is usually large and only g_{iso}, or $g_{\parallel}$ and $g_{\perp}$, lines are seen. However, there is a pronounced orbital contribution present in octahedral geometry, so that g is about 4.3. For axially symmetric systems, the two g-values usually give an average close to 4.3 but the individual values of $g_{\parallel}$, $g_{\perp}$ may lie anywhere in the range 1–8.

Tetrahedral complexes gain orbital contribution by the indirect spin–orbit coupling mechanism, but the relevant excited states are considerably lower in energy than in the octahedral case. g-values are therefore appreciably greater than 2.0. However, the spectra are not usually readily observable.

Hyperfine coupling to the ^{59}Co nucleus ($I = 7/2$) is usually seen. $A_\parallel$ is normally much greater than $A_\perp$, and may have values up to 300 mT or more. The coupling is normally greater in octahedral than tetrahedral complexes.

BIBLIOGRAPHY

General

R. S. Alger, *Electron Paramagnetic Resonance*, Interscience, New York, 1968. Good introduction to the subject, with clear explanations.

N. M. Atherton, *Electron Spin Resonance*, Ellis Horwood, Chichester, 1973 (second edition imminent).

P. W. Atkins and M. C. R. Symons, *The Structure of Inorganic Radicals*, Elsevier, Amsterdam, 1967.

P. B. Ayscough, *Electron Spin Resonance in Chemistry*, Methuen, London, 1967.

B. A. Goodman and J. B. Raynor, *Adv. Inorg. Chem. Radiochem.* (eds Emeleus and Sharpe), **13** (1970) 135. 'Electron Spin Resonance of Transition Metal Complexes'. Comprehensive coverage of all major topics, with discussions of individual metal ions.

D. J. E. Ingram, *Biological and Biochemical Applications of Electron Spin Resonance*, Hilger, London, 1969. A very readable account.

B. R. McGarvey, *Transition Metal Chemistry* (ed. Carlin), **3** (1966) 90. 'Electron Spin Resonance of Transition Metal Complexes'. Rather theoretical. Discussion of individual ions.

M. C. R. Symons, *Chemical and Biological Applications of Electron Spin Resonance Spectroscopy*, van Nostrand, New York, 1978. Quite readable; goes into considerable detail.

J. F. Wertz and J. R. Bolton, *Electron Spin Resonance*, McGraw-Hill, New York, 1986 (1st edn, 1972). Comprehensive. Good basic introduction, but rather theoretical in parts. Useful for in-depth follow-up.

Specific systems

N. D. Chasteen, *Biol. Mag. Res.* (eds Berliner and Reuben), **3** (1981) 53. 'Vanadyl(IV) EPR Spin Probes: Inorganic and Biological Aspects'. Very readable.

H. Sigel (ed.), *Metal Ions in Biological Systems* Vol. 22 (1987). 'ENDOR, EPR and Electron Spin-Echo for Probing Coordination Complexes.' Useful survey of modern methods and results (ENDOR = electron-nuclear double resonance).

PROBLEMS

Answers on pp. 197–202

P5.1 Draw the energy-level diagram (energy versus applied field) for a system with $S = 3/2$ under the various circumstances listed below. Select a particular energy interval as corresponding to the spectrometer frequency, mark the possible transitions, and note how their number changes [assume that the limit of applied field available is twice that corresponding to the transition in (a)].

(a) with no zero-field splitting,
(b) with small zero-field splitting,
(c) with zero-field splitting comparable to the EPR spectrometer frequency, and
(d) with zero-field splitting very much larger than the spectrometer frequency.

P5.2 Repeat the exercise of Problem 5.1 for $S = 2$.

P5.3 From the spectrum in Fig. 5.21, estimate the g-value.

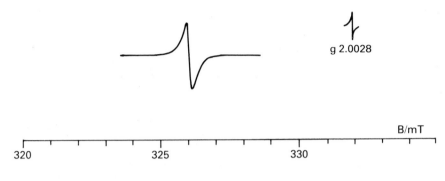

Fig. 5.21.

P5.4 From the spectrum shown in Fig. 5.22, estimate the value of g and of A. What type of spin system would give rise to this type of structure in the spectrum?

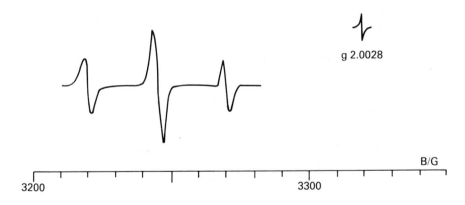

Fig. 5.22.

P5.5 Sketch the shape of the EPR spectra expected for the following species, in which M has *no* nuclear spin ($I = 0$).
(a) M˙   (b) MH˙   (c) $MH_2˙$ (d) $MH_3˙$

P5.6 Sketch the EPR spectrum expected for $AlH_3˙$ (^{27}Al, 100%, $I = 5/2$).
Compare your sketch with Fig. 5.23, which is the experimentally observed

spectrum, and estimate the magnitudes of the hyperfine and superhyperfine coupling constants in the spectrum.

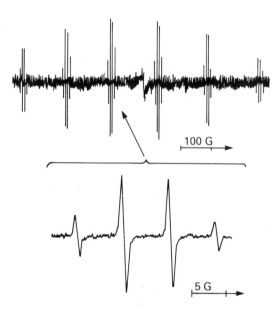

100 G

5 G

Fig. 5.23 — EPR spectrum of $AlH_3^{\cdot}$ (Reproduced with permission from J. R. M. Giles and B. P. Roberts, *J. Chem. Soc., Chem. Comm.*, (1981) 1167.)

P5.7 The spectrum shown in Fig. 5.24 is for the radical anion $B_2H_6^{\cdot-}$, in which the

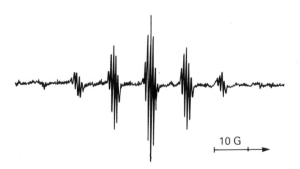

10 G

Fig. 5.24 — EPR spectrum of $B_2H_6^{\cdot-}$ (Reproduced with permission from J. R. M. Giles and B. P. Roberts, *J. Chem. Soc., Chem. Comm.*, (1984) 273.)

boron atoms have been isotopically enriched to 98% in ^{11}B ($I = 3/2$). What can be deduced about the numbers of boron and hydrogen nuclei with which the unpaired electron is interacting? What implications does this have for the structure of the radical anion?

P5.8   Explain why minerals which contain iron only in oxidation state $+2$ do not give usually an EPR spectrum.

P5.9   The spectra shown in Fig. 5.25 are of a sample of a synthetic iron silicate which has been (a) calcined (i.e. heated in air air) and (b) then extracted with oxalic acid solution. The marker signal is pitch, $g = 2.0028$. It is thought that the samples contain iron(III) both in tetrahedral sites in the polymeric silicate framework and in octahedrally coordinated interstitial sites. Identify the signals corresponding to the two sites, and explain the change in their relative intensities (marked on the spectra as figures in parentheses).

P5.10   Sketch the EPR spectra you would expect to see for

(a) a solution of Cu^{2+} in water, and
(b) the same solution after it has been frozen.

[Assume that only hyperfine coupling to ^{63}Cu ($I = 3/2$) can be seen.]

P5.11   A marine biological organism has a vanadium-containing molecule which catalyses reactions of the type

$$Br^- + RH + H_2O_2 \rightarrow RBr + OH^- + H_2O$$

In its functional form this molecule is EPR silent but it will undergo a one-electron reduction to give a species with an anisotropic EPR spectrum with $g_{||} = 1.984$, $A_{||} = 17.6\,mT$, $g_\perp = 1.979$, $A_\perp = 5.7\,mT$.
   Infer the oxidation state of vanadium in the functional form of the molecule.
   Explain the structural significance of an anisotropic EPR spectrum.
   Draw a stick diagram of the expected line positions (first-order) as a function of magnetic field for a spectrometer operating at 9.458 GHz.

$$[^{51}V, 100\%, I = 7/2; \text{ take } h/\mu = 7.145 \times 10^{-11}\,Ts]$$

ANSWERS

A5.1   See Fig. 5.26. Note that the $m_s = -1/2 \rightarrow +1/2$ transition is unaffected by zero-field splitting, and always appears, even though the other transitions may move out of the spectrometer range.

A5.2   See Fig. 5.27. Note the dependence of all the line positions on the zero-field splitting and that it is possible for there to be no lines at all in the spectrometer range.

A5.3   The centre of the line appears at about 326 mT and the reference signal at about 332 mT. Hence $g = 2.0028 \times 332/326 = 2.04$.

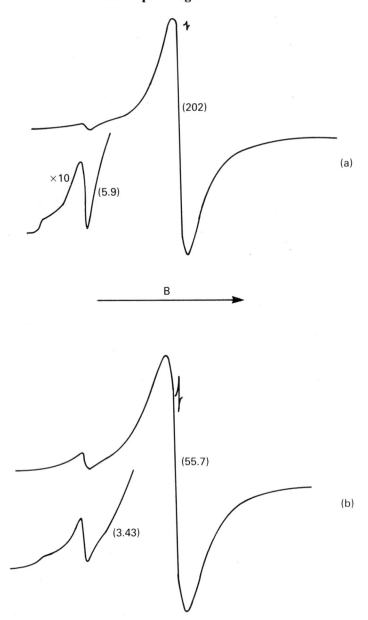

Fig. 5.25 — EPR spectrum of a powdered FAPO sample.

A5.4   The spectrum is a 1:2:1 triplet, centred at $B = 3245G$. The reference signal is at $B = 3319G$, hence $g = 2.048$. The hyperfine coupling constant is the separation between the individual lines, $25G$. Such a spectrum would arise from coupling of the unpaired electron with two nuclei with $I = 1/2$.

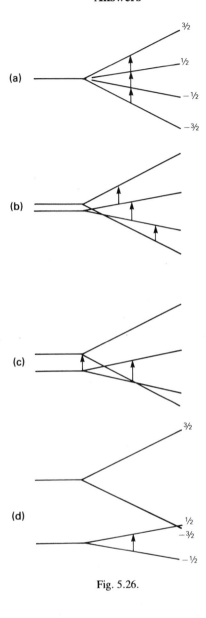

Fig. 5.26.

A5.5   Spectra should have the following shapes: (a) a single line, (b) a doublet, (c) a 1:2:1 triplet; (d) a 1:3:3:1 quartet. In (c) and (d) the lines are equally spaced.

A5.6   Coupling to ^{1}H will give quartet structure as in Problem 5.5(d). Coupling to ^{27}Al will give six equally intense lines. Since $A(^{27}$Al$) \gg A(^1$H$)$, the spectrum will appear as a sextet of quartets.

The experimental spectrum shows second-order behaviour in that the intensities of the sextet lines are not equal. Careful observation also shows that the sextet line

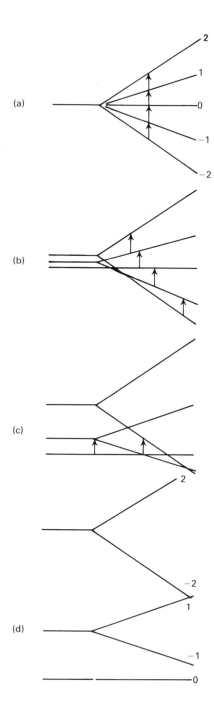

Fig. 5.27.

spacing increases slightly with increasing magnetic field, from $140G$ to $164G$. $A(^{27}Al)$ is the average of these extreme values, $152G$ ($15.2\,\text{mT}$). $A(^1H)$ is the line spacing of the quartets, $7.3G$ ($0.73\,\text{mT}$).

A5.7 The spectrum is a septet of septets. The large line-separation must be hyperfine splitting from ^{11}B. Therefore the unpaired electron is interacting equally with both boron nuclei, i.e. is delocalized between the two boron atoms. The small splitting must be due to superhyperfine interaction with all six ^1H nuclei, which must also be equivalent.

Diborane, B_2H_6, has an electron-deficient bridged structure containing four terminal and two bridging hydrogen atoms. The radical does not have this structure, since the hydrogen atoms are now all equivalent. The simplest structure would be analogous to ethane, with only a single bonding electron holding the two BH_3-units together.

A5.8 Most minerals contain high-spin iron(II) which has a d^6 configuration with four unpaired electrons ($S = 2$). Even-electron systems seldom give EPR spectra (see Problem 5.2).

A5.9 From section 5.3.2.1, Fe^{3+} in octahedral and tetrahedral coordination gives g-values of about 2.0 and about 4.3 repespectively. The more intense signal is close to the marker and therefore has $g \sim 2.0$ and corresponds to the octahedral Fe^{3+}. Treatment with oxalic acid removes more of this species than of the tetrahedrally coordinated framework iron, showing that the interstitial iron is more accessible and more reactive.

A5.10 See Fig. 5.28. (a) Solution spectra are isotropic even though the individual

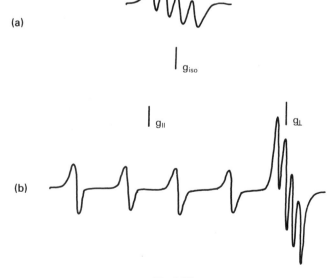

Fig. 5.28.

ions may have non-cubic symmetry (see section 5.2.3). A simple quartet spectrum is therefore expected. (b) When the solution is frozen, the ions are immobilized and have a random distribution of orientations. An anisotropic spectrum showing $g_\parallel$ and $g_\parallel$ is therefore obtained. Since $A_\parallel$ is normally quite small for Cu^{2+}, A_{iso} is roughly $A_\perp/3$.

A5.11 The readily available oxidation states for vanadium are $+5$ and $+4$. The V^{5+} ion has a d^0 configuration and gives no EPR spectra. Vanadium(IV) almost always occurs as VO^{2+} which has five other ligands coordinated in an approximate octahedral arrangement. This system would give an anisotropic EPR signal showing $g_\parallel$ and $g_\perp$.

Rearranging the resonance condition gives $B = h\nu/g\mu$, where $\nu = 9.458 \times 10^9\,s^{-1}$ and $h/\beta = 7.145 \times 10^{-11}\,Ts$. Hence the fields corresponding to $g_\parallel$ and $g_\perp$ are $0.3406\,T$ and $0.3415\,T$. Both components will be split into eight, with separations of $17.6n\,mT$ and $5.7n\,mT$ from these positions, where $n = \pm 3.5$, ± 2.5, ± 1.5 and ± 0.5. The line positions are therefore (in T).

$g_\parallel$	$g_\perp$
0.2790	0.3215
0.2966	0.3272
0.3142	0.3329
0.3319	0.3386
0.3494	0.3443
0.3761	0.3500
0.3846	0.3557
0.4022	0.3614

Note that the two sets of lines overlap (Fig. 5.29).

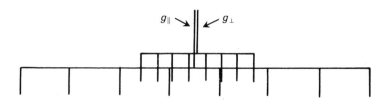

Fig. 5.29.

Appendix 1
Properties of NMR isotopes

(Those with abundance less than 1.0% have been ignored.)

^{1}H	
99.99	Isotope
1/2	Abundance/%
100.0	Spin
1.00	Relative frequency/MHz
26.75	Relative receptivity (*vs* ^{1}H)
	Magnetogyric ratio/10^7 rad T^{-1} s^{-1}

^{7}Li*	^{9}Be
92.6	100
3/2	3/2
38.9	14.1
0.27	1.4E−2
10.4	−3.76

^{23}Na	^{25}Mg
100	10.1
3/2	5/2
26.5	6.1
9.2E−2	2.7E−4
−3.63	−1.64

^{39}K	^{43}Ca	^{45}Sc	^{47}Ti*	^{51}V	^{53}Cr
93.1	0.15	100	7.3	99.8	9.6
3/2	7/2	7/2	5/2	7/2	3/2
4.7	6.7	24.3	5.6	26.3	5.7
4.7E−4	9.3E−6	0.30	1.5E−4	0.38	8.6E−5
1.25	−1.80	6.51	−1.51	7.05	−1.5

^{87}Rb*	^{87}Sr	^{89}Y	^{91}Zr	^{93}Nb	^{95}Mo*
27.9	7.0	100.0	11.2	100	15.7
3/2	9/2	1/2	5/2	9/2	5/2
32.8	4.3	4.9	9.3	24.5	6.5
4.9E−2	1.9E−4	1.2E−4	1.1E−3	0.49	5.1E−4
8.78	−1.63	−1.32	−2.50	6.56	1.75

^{133}Cs	^{137}Ba*	^{139}La	^{177}Hf*	^{181}Ta	^{183}W
100	11.3	99.9	18.5	100	14.4
7/2	3/2	7/2	7/2	7/2	1/2
13.2	11.1	14.2	4.0	12.0	4.2
4.8E−2	7.8E−4	6.0E−2	2.0E−4	3.6E−2	1.0E−5
3.53	2.99	3.80	1.08	3.22	1.12

* Other isotopes

^{6}Li	^{49}Ti	^{85}Rb	^{97}Mo	^{135}Ba	^{179}Hf
7.4	5.5	72.8	9.5	6.6	13.8
1	7/2	5/2	5/2	3/2	9/2
14.7	6.4	9.7	6.7	9.9	2.5
6.3E−4	2.1E−4	7.6E−3	3.3E−4	3.3E−4	6.5E−5
3.94	−1.51	2.59	−1.79	2.68	−0.68

Isotope
Abundance/%
Spin
Relative frequency/MHz
Relative receptivity (vs ^{1}H)
Magnetogyric ratio/10^7 rad T^{-1} s^{-1}

^{55}Mn	^{57}Fe	^{59}Co	^{61}Ni	^{63}Cu*	^{67}Zn
100	2.2	100	1.2	69	4.1
5/2	1/2	7/2	3/2	3/2	5/2
24.8	3.24	23.7	8.93	26.5	6.26
0.18	7.5E−7	0.28	4.3E−5	6.5E−2	1.2E−4
6.64	0.87	6.30	−2.39	7.11	1.68
	^{101}Ru*	^{103}Rh	^{105}Pd	^{109}Ag*	^{113}Cd*
	17.1	100	22.2	48.2	12.3
	5/2	1/2	5/2	1/2	1/2
	5.2	3.2	4.6	4.7	22.2
	2.7E−4	3.1E−5	2.5E−4	4.9E−5	1.3E−3
	−1.38	−0.85	−1.23	−1.25	−5.70
^{187}Re*	^{187}Os	^{193}Ir*	^{195}Pt	^{197}Au	^{199}Hg*
62.9	1.6	62.7	33.8	100	16.8
5/2	1/2	3/2	1/2	3/2	1/2
23.1	2.3	1.9	21.4	1.75	17.9
9.0E−2	2.0E−7	9.9E−5	3.5E−3	2.5E−5	1.0E−3
6.17	0.62	0.51	5.84	0.47	4.85

* Other isotopes

^{65}Cu	^{99}Ru	^{107}Ag	^{111}Cd	^{185}Re	^{191}Ir	^{201}Hg
30.9	12.7	51.8	5	37.1	37.3	13.2
3/2	5/2	1/2	5/2	5/2	3/2	3/2
28.4	4.6	4.0	6.7	22.8	1.74	6.61
3.6E−2	1.4E−4	3.5E−5	1.2E−3	5.0E−2	7E−6	2.0E−4
7.61	−1.23	−1.09	−1.79	6.11	0.47	−1.79

^{189}Os
16.1
3/2
7.77
3.8E−4
2.08

^{11}B*	^{13}C	^{14}N*		^{19}F	
80.4	1.11	99.6		100	
3/2	1/2	1		1/2	
32.1	25.1	7.23		94.1	
0.13	1.8E−4	1.0E−3		0.83	
8.58	6.73	1.93		25.2	
^{27}Al	^{29}Si	^{31}P		^{35}Cl*	
100	4.7	100		75.5	
5/2	1/2	1/2		3/2	
26.1	19.1	40.5		9.80	
0.21	3.7E−4	6.6E−2		3.6E−3	
6.97	−5.31	10.8		2.62	
^{69}Ga*	^{73}Ge	^{75}As	^{77}Se	^{81}Br*	^{83}Kr
60.4	7.8	100	7.6	49.5	11.6
3/2	9/2	3/2	1/2	3/2	9/2
24.0	3.5	17.2	19.1	27.1	3.9
4.0E−2	1.1E−4	2.5E−2	5.3E−4	4.9E−2	2.2E−4
6.70	−0.94	4.60	5.12	7.25	−1.03
^{115}In*	^{119}Sn*	^{121}Sb*	^{125}Te*	^{127}I	^{129}Xe*
95.7	8.6	57.3	7.0	100	26.4
9/2	1/2	5/2	1/2	5/2	1/2
22.0	37.3	24.1	31.5	20.1	27.9
0.33	4.5E−3	9.3E−2	2.2E−3	9.5E−2	5.7E−3
5.90	−10.00	6.44	−8.51	5.39	−7.45
^{205}Tl	^{207}Pb	^{209}Bi			
70.5	22.6	100			
1/2	1/2	9/2			
57.6	20.9	16.4			
0.14	2.1E−3	0.14			
15.7	5.62	4.38			

* Other isotopes

^{10}B	^{15}N	^{37}Cl	^{71}Ga	^{79}Br	^{113}In	^{117}Sn
19.6	0.37	24.5	39.6	50.4	4.3	7.6
3	1/2	3/2	3/2	3/2	9/2	1/2
10.7	10.1	8.16	30.6	25.1	22.0	35.6
3.9E−3	3.9E−6	6.7E−4	5.7E−2	4.0E−2	1.5E−2	3.5E−3
2.87	−2.71	2.18	8.18	6.70	5.89	−9.59

^{123}Sb	^{123}Te	^{131}Xe	^{203}Tl
42.8	0.87	21.2	29.5
7/2	1/2	3/2	1/2
13.0	26.2	8.26	57.1
2.0E−2	1.6E−4	6.0E−4	5.4E−2
3.49	−7.06	2.21	15.5

Lanthanides	Isotope Abundance/% Spin Relative frequency/MHz Relative receptivity (vs ^1H) Magnetogyric ratio/10^7 rad T^{-1} s^{-1}					

	^{141}Pr	^{143}Nd*		^{147}Sm*	^{153}Eu*	^{157}Gd*
	100	12.2		15.0	52.2	15.7
	5/2	7/2		7/2	5/2	3/2
	29.0	5.51		4.16	11.0	4.03
	0.29	4.2E−4		2.3E−4	8.0E−3	4.5E−5
	7.77	−1.47		−1.11	2.94	−1.08

^{159}Tb	^{163}Dy*	^{165}Ho	^{167}Er	^{169}Tm	^{171}Yb*	^{175}Lu*
100	25.0	100	22.9	100	14.3	97.4
3/2	5/2	7/2	7/2	1/2	1/2	7/2
24.0	4.77	21.3	2.90	7.99	17.7	11.4
6.5E−2	3.0E−4	0.20	1.2E−4	5.3E−4	7.5E−4	3.0E−2
6.43	1.28	5.71	−0.78	−2.14	4.73	3.06

* Other isotopes

^{145}Nd	^{149}Sm	^{151}Eu	^{155}Gd	^{161}Dy	^{173}Yb	^{176}Lu
8.3	13.8	47.8	14.7	18.9	16.1	2.6
7/2	7/2	5/2	3/2	5/2	5/2	7
3.41	3.43	24.5	3.09	3.44	4.87	8.10
6.8E−5	1.1E−4	8.3E−2	2.2E−5	8.5E−5	2.1E−4	1.0E−4
−0.91	−0.92	6.55	−0.83	−0.92	−1.30	2.17

Adapted from the compilations in J. Mason (ed.), *Multinuclear NMR*, Plenum Press, New York, 1987, and J. B. Lambert and F. G. Riddell (eds.), *The Multinuclear Approach to NMR Spectroscopy*, D. Reidel Publishing Co., Dordrecht (Netherlands), 1983.

Appendix 2
Properties of NQR isotopes

(Those with abundance less than 1.0% have been ignored.)

Note: There is much uncertainty over the precise numerical values of nuclear quadrupole moments (see below).

		Isotope Abundance/% Spin Quadrupole moment/10^{-28} m^2			
^{7}Li* 92.6 3/2 -4E-2	^{9}Be 100 3/2 5E-2				
^{23}Na 100 3/2 0.1	^{25}Mg 10.1 5/2 0.2				
^{39}K* 93.1 3/2 5E-2		^{45}Sc 100 7/2 -0.2	^{47}Ti* 7.3 5/2 0.3	^{51}V 99.89 7/2 -5E-2	^{53}Cr 9.6 3/2 -0.2
^{85}Rb* 72.2 5/2 0.3	^{87}Sr 7.0 9/2 0.4		^{91}Zr 11.2 5/2 -0.2	^{93}Nb 100 9/2 -0.2	^{95}Mo* 15.7 5/2 -0.1
^{133}Cs 100 7/2 -3E-3	^{137}Ba* 11.3 3/2 0.3	^{139}La 99.9 7/2 0.2	^{177}Hf* 18.5 7/2 4	^{181}Ta 100 7/2 3	

* Other isotopes

^{6}Li 7.4 1 -8E-4	^{41}K 6.9 3/2 7E-2	^{49}Ti 5.5 7/2 0.2	^{87}Rb 27.8 3/2 0.1	^{97}Mo 9.5 5/2 0.2	^{135}Ba 6.6 3/2 0.2	^{179}Hf 13.8 9/2 4

| Isotope |
| Abundance/% |
| Spin |
| Quadrupole moment/10^{-28} m^2 |

^{55}Mn		^{59}Co	^{61}Ni	^{63}Cu*	^{67}Zn
100		100	1.2	69.1	4.1
5/2		7/2	3/2	3/2	3/2
0.5		0.4	0.2	−0.2	−0.2
	^{101}Ru*		^{105}Pd		
	17.1		22.2		
	5/2		5/2		
	0.4		0.8		
^{187}Re*	^{189}Os	^{193}Ir*		^{197}Au	^{201}Hg
62.9	16.1	62.7		100	13.2
5/2	3/2	3/2		3/2	3/2
2	1	1		0.6	0.5

* Other isotopes

^{65}Cu	^{99}Ru	^{185}Re	^{191}Ir
30.9	12.7	37.1	37.3
3/2	5/2	5/2	3/2
−0.2	7E−2	2	1

¹¹B* 80.4 3/2 3E−2		¹⁴N* 99.6 1 2E−2		²¹Ne 0.26 3/2 8E−2
²⁷Al 100 5/2 0.2			³⁵Cl* 75.5 3/2 −8E−2	
⁶⁹Ga* 60.4 3/2 0.2	⁷³Ge 7.8 9/2 −0.2	⁷⁵As 100 3/2 0.3	⁷⁹Br* 50.5 3/2 0.3	⁸³Kr 11.6 9/2 0.2
¹¹⁵In* 95.7 9/2 0.8		¹²¹Sb* 57.3 5/2 −0.5	¹²⁷I 100 5/2 −0.8	¹³¹Xe 21.2 3/2 −0.1
		²⁰⁹Bi 100 9/2 −0.4		

* Other isotopes

¹⁰B	³⁷Cl	⁷¹Ga	⁸¹Br	¹¹³In	¹²³Sb
19.6	24.5	39.6	49.5	4.3	42.8
3	3/2	3/2	3/2	9/2	7/2
7E−2	−6E−2	0.1	0.3	1	−0.7

				Isotope Abundance/% Spin Quadrupole moment/10^{28} m^2		

Lanthanides

	^{141}Pr	^{143}Nd*		^{147}Sm*	^{153}Eu*	^{157}Gd*
	100	12.2		15.0	52.2	15.7
	5/2	7/2		7/2	5/2	3/2
	5E−2	−0.5		−0.2	3	2

^{159}Tb	^{163}Dy*	^{165}Ho	^{167}Er		^{173}Yb	^{175}Lu*
100	25.0	100	22.9		16.1	97.4
3/2	5/2	7/2	7/2		5/2	7/2
1	2	3	3		3	4

* Other isotopes

^{145}Nd	^{149}Sm	^{151}Eu	^{155}Gd	^{161}Dy	^{176}Lu
8.3	13.8	47.8	14.7	18.9	2.6
7/2	7/2	5/2	3/2	5/2	7
−0.2	6E−2	1	2	2	8

These data have been collected from various sources. Nuclear quadrupole moments can only be evaluated from a knowledge of the EFG for a system for which the QCC has been measured. Two methods have been adopted: (a) from assumed electron configurations for particular systems and estimates of the electronic radial functions, or (b) from molecular-orbital calculations to give electron distributions. Unfortunately, the assumed electronic configurations have often been very unrealistic, radial functions for electrons in molecules are not accurately known and the wavefunctions used in molecular-orbital calculations are not very good in the region of the nucleus.

Appendix 3
Isotopes for Mössbauer spectroscopy

Legend:

Isotope
Abundance/%
E_γ/keV
I_{gd}
I_{ex}
Γ_{nat}/mm s^{-1}

^{40}K					
1.2E$-$2					
29.4					
4					
3					
2.2					

^{133}Cs	^{133}Ba	^{139}La*	^{180}Hf*	^{181}Ta	^{182}W*
100	0	99.9	35.2	99.9	26.4
81.0	12.3	166	93.3	6.25	100
7/2	1/2	7/2	0	7/2	0
5/2	3/2	5/2	2	9/2	2
0.54	2.7	1.1	2.0	6.4E$-$3	2.1

* Other isotopes are also possible

VII	VIII	IX	X	XI	XII
	^{57}Fe		^{61}Ni		^{67}Zn
	2.2		1.2		4.1
	14.4		67.4		93.3
	1/2		3/2		5/2
	3/2		5/2		1/2
	0.19		0.77		3.2E−4
^{99}Tc	^{99}Ru				
0	12.7				
141	89.4				
9/2	5/2				
7/2	3/2				
8.2	0.15				
^{187}Re	^{189}Os*	^{193}Ir*	^{195}Pt	^{197}Au	^{201}Hg*
62.9	16.1	62.7	33.8	100	13.2
134	36.2	73.0	98.8	77.3	32.2
5/2	3/2	3/2	1/2	3/2	3/2
7/2	1/2	1/2	3/2	1/2	1/2
204	15.1	0.59	16.3	1.9	42

* Other isotopes are also possible

| Isotope |
| Abundance/% |
| E_γ/ke V |
| I_{gd} |
| I_{ex} |
| Γ_{nat}/mm s^{-1} |

	^{73}Ge 7.8 67.0 9/2 5/2 7E-3				^{83}Kr 11.5 9.4 9/2 7/2 0.20
	^{119}Sn* 8.6 23.9 1/2 3/2 0.65	^{121}Sb 57.3 37.2 5/2 7/2 2.1	^{125}Te 7.14 35.5 1/2 3/2 5.2	^{127}I* 100 57.6 5/2 7/2 2.5	^{129}Xe* 26.4 39.6 1/2 3/2 6.8

* Other isotopes are also possible

Lanthanides

	^{141}Pr*	^{145}Nd	^{145}Pm*	^{154}Sm*	^{151}Eu*
	100	8.3	0	22.7	47.5
	145	67.3	61.2	82.0	21.6
	5/2	7/2	5/2	0	5/2
	7/2	3/2	7/2	2	7/2
	1.0	0.14	1.7	1.1	1.3

^{158}Gd*	^{159}Tb	^{161}Dy*	^{165}Ho	^{166}Er*	^{169}Tm
24.9	100	18.9	100	33.2	100
79.5	58.0	25.7	94.7	80.6	8.4
0	3/2	5/2	7/2	0	1/2
2	5/2	5/2	9/2	2	3/2
1.4	45	0.38	130	1.8	8.1

^{171}Yb	^{175}Lu
14.3	97.4
66.7	114
1/2	7/2
3/2	9/2
4.7	24

Actinides

^{232}Th	^{231}Pa	^{238}U*	^{237}Np	^{239}Pu	^{243}Am
0	0	99.3	0	0	0
49.4	84.2	44.7	59.5	57.2	83.9
0	3/2	0	5/2	1/2	5/2
2	5/2	2	5/2	5/2	5/2
16	7.9E−2	27	6.7E−2	47	1.4

*Other isotopes are also possible

From the tabulation of the Mössbauer Effect Data Center, University of North Carolina, NC 28804–3299, USA.

Index of compounds and species

General index